HarmonyOS
物联网开发基础实践

葛非◎编著

清华大学出版社
北京

内 容 简 介

本书内容丰富，涵盖 HarmonyOS 物联网开发技术方面的基础实验，包括 LiteOS 微内核基础功能实验、轻量级系统设备开发实验和应用 UI 开发实验，涉及物联网操作系统原理、海思 RISC-V CPU 接口和传感器应用等硬件技术、JavaScript 和 eTS 等 Web 前端开发技术和手机 App 开发技术、WiFi 网络应用、WebSocket 和 MQTT 等网络协议的内容。

本书共 4 章。第 1 章介绍实验环境，包含 LiteOS Studio、DevEco Device Tool 和 DevEco Studio 等开发工具、Qemu 模拟器和环境配置。第 2 章讲解 LiteOS 微内核的基本功能实验，包括任务管理、内存管理、任务间通信和软件定时器等。第 3 章学习轻量级系统设备开发中的 GPIO 输入输出、I2C 接口、PWM 接口、WiFi 连接和 MQTT 客户端的实验。第 4 章内容包含 HarmonyOS 系统应用 UI 开发技术中的组件实验和应用 JavaScript、eTS 等语言开发 App 等实验。

本书适合作为高等学校物联网、计算机专业的本科生教程，也可作为对 HarmonyOS 感兴趣的开发人员、广大科技工作者和研究人员的参考用书。

本书封面贴有清华大学出版社防伪标签，无标签者不得销售。
版权所有，侵权必究。举报: 010-62782989, beiqinquan@tup.tsinghua.edu.cn。

图书在版编目(CIP)数据

HarmonyOS 物联网开发基础实践/葛非编著. —北京: 清华大学出版社, 2023.2(2024.6 重印)
ISBN 978-7-302-62630-5

Ⅰ. ①H… Ⅱ. ①葛… Ⅲ. ①物联网—程序设计 Ⅳ. ①TP393.4 ②TP18

中国国家版本馆 CIP 数据核字(2023)第 019662 号

责任编辑: 安 妮 薛 阳
封面设计: 刘 键
责任校对: 徐俊伟
责任印制: 沈 露

出版发行: 清华大学出版社
网　　址: https://www.tup.com.cn, https://www.wqxuetang.com
地　　址: 北京清华大学学研大厦 A 座　　邮　编: 100084
社 总 机: 010-83470000　　邮　购: 010-62786544
投稿与读者服务: 010-62776969, c-service@tup.tsinghua.edu.cn
质量反馈: 010-62772015, zhiliang@tup.tsinghua.edu.cn
课件下载: https://www.tup.com.cn, 010-83470236

印 装 者: 大厂回族自治县彩虹印刷有限公司
经　　销: 全国新华书店
开　　本: 185mm×260mm　　印　张: 14.5　　字　数: 355 千字
版　　次: 2023 年 2 月第 1 版　　印　次: 2024 年 6 月第 3 次印刷
印　　数: 2001~2500
定　　价: 59.00 元

产品编号: 097266-01

前言

2020年,华为终端有限公司正式发布了鸿蒙HarmonyOS 1.0。2021年又先后正式发布了HarmonyOS 2.0、HarmonyOS 3.0和HarmonyOS 3.1 Beta。HarmonyOS系统是面向万物互联的全场景分布式操作系统,支持智能手机、平板电脑、智能穿戴设备、智慧屏和车机等多种终端设备。为不同设备的智能化、互联和协同提供了统一的语言,为用户带来简洁、流畅、安全、连续、安全可靠的全场景交互体验。HarmonyOS源代码在发布时同时开源,开源版本称为OpenHarmony,由开放原子开源基金会(OpenAtom Foundation)孵化及运营。

相对于Android、嵌入式Linux等系统,HarmonyOS不仅是一个手机或某一设备的单一系统,还是一个可将所有设备串联在一起的通用性系统。同时,HarmonyOS通过SDK、源代码、开发板/模组和开发工具等共同构成了完备的开发平台与工具链。这些特性使得HarmonyOS在物联网系统中具有强大的优势。

自HarmonyOS 1.0发布以后,我在所承担的物联网相关课程中,引入了在ARM架构CPU上运行的Harmony微内核系统LiteOS,以及由JavaScript开发运行于智能手表用户界面(UI)等相关知识内容,受到学生的欢迎。在教学过程中遇到的问题非常多,其中之一是难以找到适合的参考书籍。虽然在华为的开发者社区、HiHope开发者社区、51CTO等网站存在诸多的文档和代码,但是这些资料仍旧需要重新整理才能适应教学和学习的需要。

本书希望在HarmonyOS物联网开发技术方面为初学者提供必要的支持,因此在内容上涵盖了微内核、设备开发和应用开发的基础内容,通过这些内容可以学习LiteOS内核、轻量系统设备开发、应用开发的UI开发等基础技术。本书以实践为主,包括实验环境、LiteOS实验、轻量级系统设备开发实验和应用UI开发实验等内容,共约50个实验。内容涉及操作系统原理、接口和传感器等硬件和前端开发技术,WebSocket、MQTT等网络协议方面也有所介绍。附录中包含WiFi IoT核心板Hi3861的GPIO配置和GPIO扩展功能源代码,以及构建系统的简介。

在学习成本上,希望初学者能够零成本接触和入门HarmonyOS系统内核,低成本学习轻量级系统设备开发,零成本深入理解应用UI开发。除了设备开发的实验内容需要实物硬件设备的支持,其他实验均可通过本地模拟器、远程模拟器等方式进行,并且设备开发的大部分实验仅需要一块核心板即可完成。

读者也可以根据自己的实际情况进行取舍或减裁。如对LiteOS微内核有兴趣,可练习第2章的实验;如对物联网设备开发感兴趣,可练习第3章的实验;如对智能手机、智慧屏、车机等设备的UI应用开发有兴趣,可练习第4章的实验。完成第2章的实验需要掌握一定的操作系统原理、C程序设计、数据结构以及计算机体系结构的知识,完成第3章的实验需要掌握单片机原理、接口技术、C 程序设计、编译构建、嵌入式系统和网络协议等知识,完成第 4 章的实验仅需要掌握一定的程序设计基础知识。

本书适合作为高等学校物联网、计算机专业的本科生教材,也可作为对 HarmonyOS 感兴趣的开发人员、广大科技工作者和研究人员的参考用书。

本书在编写过程中得到教育部产学合作协同育人项目、华中师范大学-华为"智能基座"产教融合协同育人基地、华为技术有限公司和武汉科云信息技术有限公司的大力支持,在此一并表示衷心的感谢。

由于作者水平有限,书中不足之处在所难免,欢迎广大同行和读者批评指正。

葛 非

2022 年 10 月

目 录

第1章 实验环境 ··· 1

- 1.1 硬件环境 ·· 1
  - 1.1.1 Qemu 模拟器 ·· 1
  - 1.1.2 Hi3861 WLAN 模组 ··· 3
  - 1.1.3 应用运行环境 ·· 3
- 1.2 软件工具 ·· 4
  - 1.2.1 LiteOS Studio ·· 4
  - 1.2.2 STM32 工程工具 ·· 6
  - 1.2.3 Hi3861V100 WiFi IoT 工程工具 ·· 7
  - 1.2.4 DevEco Device Tool ·· 8
- 1.3 LiteOS Studio 实验环境 ·· 11
  - 1.3.1 LiteOS Studio 工程配置 ··· 11
  - 1.3.2 STM32 工程环境 ·· 15
  - 1.3.3 LiteOS shell ··· 19
- 1.4 DevEco Device Tool 实验环境 ·· 20
  - 1.4.1 Windows 系统下编译 ··· 20
  - 1.4.2 Docker 环境下编译 ··· 22
  - 1.4.3 烧录 ··· 25
  - 1.4.4 调试 ··· 25
- 1.5 DevEco Studio 实验环境 ·· 25

第2章 LiteOS 实验 ··· 26

- 2.1 LiteOS 实验概览 ··· 26
- 2.2 LiteOS 代码目录 ··· 27
- 2.3 创建任务实验 ·· 32
- 2.4 多核任务创建实验 ··· 37
- 2.5 调试任务实验 ·· 40
- 2.6 动态内存分配实验 ··· 42
- 2.7 静态内存分配实验 ··· 43
- 2.8 中断处理和错误处理实验 ·· 45

2.9	消息队列实验	47
2.10	事件实验	50
2.11	信号量实验	52
2.12	互斥锁实验	55
2.13	自旋锁实验	59
2.14	时间转换实验	62
2.15	软件定时器实验	63
2.16	注册 shell 命令实验	65
2.17	死锁发现实验	66
2.18	调度统计实验	70
2.19	CPU 利用率实验	71

第 3 章 轻量级系统设备开发实验 … 74

3.1	轻量级系统设备开发实验概览	74
3.2	Hi3861 GPIO 输出实验	75
3.3	Hi3861 GPIO 查询方式输入实验	78
3.4	Hi3861 GPIO 中断方式输入实验	81
3.5	Hi3861 PWM 输出实验	84
3.6	Hi3861 I2C 读取 AHT 实验	88
3.7	Hi3861 AT 指令实验	94
3.8	Hi3861 WiFi 连接实验	97
3.9	Hi3861 MQTT 客户端实验	104

第 4 章 应用 UI 开发实验 … 112

4.1	应用 UI 开发实验概览	112
4.2	类 Web 开发 UI 组件 Input 实验	113
4.3	类 Web 开发 UI 组件 Button 实验	115
4.4	类 Web 开发 UI 组件 Form 实验	119
4.5	类 Web 开发 UI 组件 Image 实验	121
4.6	类 Web 开发 UI 组件 Picker 实验	123
4.7	类 Web 开发 UI 组件 Tabs 实验	127
4.8	页面路由实验	129
4.9	js2java-codegen 工具应用实验	132
4.10	类 Web 开发 UI 实验	135
4.11	声明式开发 UI 组件 Button 实验	141
4.12	声明式开发 UI 组件 Text 实验	145
4.13	声明式开发 UI 组件 Image 实验	148
4.14	声明式开发 UI 组件 Slider 实验	155
4.15	声明式开发 UI 组件 Flex 实验	157

4.16	声明式开发 UI 组件 Stack 实验	165
4.17	声明式开发 UI 组件 Tabs	166
4.18	声明式开发 UI 组件 List 实验	168
4.19	声明式开发 UI 组件 Grid 实验	170
4.20	声明式开发 UI 自定义组件实验	174
4.21	声明式开发多组件 UI 实验	179
4.22	WebSocket 客户端实验	192
4.23	MQTT 客户端实验	194

附录 A WiFi IoT 核心板 GPIO 配置 — 197

附录 B GPIO 扩展功能源代码文件 — 200

- B.1 wifiiot_gpio_ex.h — 200
- B.2 wifiiot_gpio_ex.c — 208
- B.3 BUILD.gn — 208

附录 C 系统编译与构建 — 210

- C.1 Ninja 系统 — 210
- C.2 gn 系统 — 212
- C.3 轻量级系统编译构建 — 219

后记 — 223

第1章

实验环境

> 什么是路？就是从没路的地方践踏出来的，从没有荆棘的地方开辟出来的。
> 以前早有路了，以后也该永远有路。
>
> ——鲁迅

1.1 硬件环境

应用 OpenHarmony 等系统进行物联网开发通常需要硬件的支持。在初学时应尽量多地通过模拟器进行。因此除非必要，本书的实验均在模拟器或仿真器上进行。

1.1.1 Qemu 模拟器

LiteOS 支持的硬件架构包括 ARM、RISC-V 和 C-SKY 等。具体有 ARM 的 Cortex-M0、Cortex-M0+、Cortex-M3、Cortex-M4、Cortex-M7、Cortex-A7、Cortex-A9、Cortex-A53 以及 ARM64 架构的 Cortex-A72，RISC-V 的 RV32 以及 C-SKY 的 CK802 架构。开源 LiteOS 工程支持 STM32F429IG、STM32L431RC、STM32F769NI、STM32F072RB、STM32F103ZE、STM32F407ZG 等 STM32 系列开发板，以及支持 Qemu 仿真的 realview-pbx-a9 开发板，Hi3861V100 开发板等。

Qemu 是一套由 Fabrice Bellard 编写的采用 GPL 许可证的模拟处理器，在 GNU/Linux 平台上使用广泛，也支持在 x86 环境上运行。Qemu 是一款通用的开源虚拟化模拟器，通过软件模拟硬件设备。当 Qemu 直接模拟 CPU 时，它能够独立运行操作系统。realview-pbx-a9 工程就是使用 Qemu 模拟了 Cortex-A9 处理器。如果开发板使用 Qemu 模拟器，则应根据情况安装 Qemu 软件。在 Windows 64 位系统下，可访问 Qemu 官网 https://qemu.weilnetz.de/w64/下载最新版本。

本书的例子中，把 Qemu 模拟器安装在 D:/Program Files/qemu 目录下。在安装过程中，务必勾选所需要模拟硬件架构的选项。如选 ARM，则在该目录下有名称为 qemu-system-arm.exe 的可执行文件。同样地，qemu-system-i386.exe 用于模拟 i386 架构的硬件。同时建议将 Qemu 所在目录加入 PATH 环境变量。

使用 Qemu 模拟器之前需要测试其是否能够正常运行，以及了解模拟器的使用。测试需要先获取某个系统的 img 镜像，以简单的单用户 DOS 操作系统为例，假如已经下载了一个文件名为 MSDOS71B.IMG 的 DOS 系统镜像，把它保存在 Qemu 目录下。在命令行中，输入以下命令：

```
qemu-system-i386 -m 8 -k en-us -rtc base=localtime -soundhw sb16,adlib
-device cirrus-vga -fda MSDOS71B.IMG
```

该命令表示使用 i386 启动 Qemu，内存设置 8MB，使用 US 布局键盘，设置虚拟 RTC 匹配本地时间，设置 SoundBlaster16 声卡和 AdLib 音乐，使用 VGA 视频卡，指定 MSDOS71B.IMG 软盘。该命令执行后，得到如图 1.1 所示的 DOS 系统。

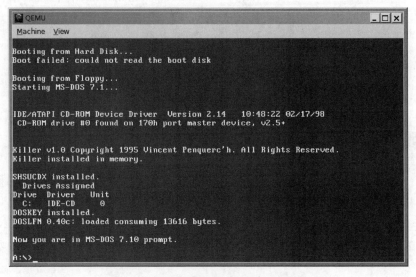

图 1.1　Qemu 仿真 i386 运行 DOS 系统

qemu-system-i386 的参数含义如下。

-m 16：定义模拟器内存。DOS 系统不需要很多的内存，此处为 16MB。

-k en-us：指定键盘布局。-k 选项是不必要的，虚拟键盘匹配真实键盘。此处指 US 布局、英语。

-rtc base=localtime：设置虚拟 RTC 匹配本地时间。每个传统的 PC 设备有一个实时时钟（RTC）以便于系统可以保持跟踪时间。

-soundhw sb16,adlib,pcspk：定义 Qemu 支持 SoundBlaster16 声卡和 AdLib 音乐。SoundBlaster16 和 AdLib 是在 DOS 时期常见的音频硬件。

-device cirrus-vga：定义 VGA 视频卡。Cirrus VGA 卡是当时比较常见的图形卡。

-display gtk：设置虚拟显示 GTK toolkit，它可以将虚拟系统放到自身的窗口内，并且提供一个简单的菜单去控制虚拟机。

-hda：指定硬盘。

-fda：指定软盘。

-cdrom：指定光盘。

-boot order=d：指定 D 盘为（DOS 下一般会分配给光盘）引导。可以让 Qemu 从多个引导源来引导虚拟机。从软盘驱动器引导(在 DOS 机器中一般情况下是 A:)指定 order＝a；从第一个硬盘驱动器引导(一般称为 C:)使用 order＝c；从 CD-ROM 驱动器引导(在 DOS 中经常分配为 D:)使用 order＝d。可以使用组合字母去指定一个特定的引导顺序，比如 order＝dc，则会先尝试使用 CD-ROM 驱动器，如果 CD-ROM 驱动器中没有引导介质，再用硬盘驱动器。

1.1.2 Hi3861 WLAN 模组

Hi3861 芯片是一款高度集成的 2.4GHz WLAN SoC 芯片，集成 IEEE 802.11b/g/n 基带和 RF 电路，Hi3861 WLAN 模组主板芯片为 Hi3861。该主板是 HiSpark WiFi-IoT 智能家居物联网开发套件或者 WiFi-IoT 智能小车套件的核心板。本书大部分的设备开发实验可以在 Hi3861 核心板上完成，部分实验需要物联网开发套件。

1.1.3 应用运行环境

HarmonyOS 的应用开发运行环境可以使用模拟器、远程真机或者本地真机。对于 Phone，可以使用 Local Emulator 和 Remote Emulator；对于 Tablet、TV 和 Wearable，可以使用 Remote Emulator 运行应用；对于 Lite Wearable 和 Smart Vision，可以使用 Simulator 运行应用。Remote Emulator 还提供超级终端模拟器(super device)，可以利用超级终端模拟器来调测跨设备应用。

Local Emulator 基于 x86 架构，可以运行和调试智能手机、TV 和智能可穿戴设备的 HarmonyOS 应用。在 Local Emulator 上运行应用兼容签名与不签名两种类型的 HAP。

Local Emulator 是在本地计算机上创建和运行的，相比于 Remote Emulator，不需要登录授权，在运行和调试应用时，由于没有网络数据的交换，因此可以保持很好的流畅性和稳定性；但是需要耗费一定的计算机磁盘资源，Windows 系统内存推荐为 16GB 及以上，mac OS 系统内存推荐为 8GB 及以上。不支持在虚拟机系统上运行本地模拟器，例如，不支持在 Ubuntu 系统上，通过安装 Windows 虚拟机，然后使用 Windows 系统安装和运行模拟器。

Remote Emulator 中的单设备模拟器(Single Device)可以运行和调试智能手机(折叠屏 Mate X2、P40 和 P40 Pro)、平板电脑(MatePad Pro)、TV 和智能可穿戴设备的 HarmonyOS 应用，可兼容签名与不签名两种类型的 HAP。

Remote Emulator 使用时长每次为 1h，到期后会自动释放资源，需要及时完成 HarmonyOS 应用的调试。不过如果 Remote Emulator 到期释放后，可以重新申请资源。

目前超级终端模拟器支持手机＋手机、手机＋平板电脑和手机＋TV 的设备组网方式，可以使用该超级终端模拟器来调测具备跨设备特性的应用，如应用在不同设备间的流转。

远程真机设备(Remote Device)资源也可供开发者使用，减少开发成本。目前，远程真机支持手机和可穿戴设备，使用远程真机调试和运行应用时，同本地物理真机设备一样，需要对应用进行签名才能运行。

相比远程模拟器，远程真机是部署在云端的真机设备资源，远程真机的界面渲染和操作体验更加流畅,同时也可以更好地验证应用在真机设备上的运行效果,比如性能、手机网络环境等。

1.2 软件工具

1.2.1 LiteOS Studio

LiteOS Studio 是以 Visual Studio Code 的社区开源代码为基础,根据 C 语言的特点和 LiteOS 嵌入式系统软件的业务场景,定制开发的一款轻量级集成开发环境,提供了代码编辑、编译、烧录和调试等功能。

在 https://gitee.com/LiteOS/LiteOS_Studio/releases 可以获取 LiteOS Studio 安装包 HUAWEI-LiteOS-Studio-Setup-x64-×.××.×,其中,×.××.× 为 LiteOS Studio 版本号,本书的例子使用的是 HUAWEI-LiteOS-Studio-Setup-x64-1.45.6 版。

除此之外,LiteOS Studio 还提供了基于 Visual Studio Code 的扩展,可供用户作为插件安装在 Visual Studio Code 或其他基于 VSCode 的定制 IDE 上。

LiteOS Studio 安装过程十分简单,双击下载的安装包 HUAWEI-LiteOS-Studio-Setup-x64-1.45.6.exe,依照软件提示,设置安装目录等选项,如图 1.2 所示。安装准备就绪后,单击"安装"按钮即可安装 LiteOS Studio 集成开发环境。

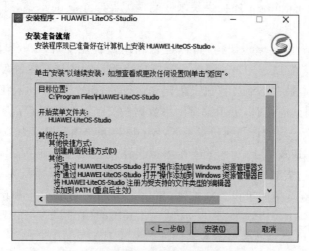

图 1.2 LiteOS 安装准备就绪对话框

安装完成后,LiteOS Studio 主界面如图 1.3 所示。LiteOS Studio 主界面的区域 1 是菜单栏;区域 2 是侧边栏;区域 3 是工程树,由项目工程文件构成,可进行快速新建及打开文件等操作;区域 4 是调试工具栏,可进行编译、烧录、调试等操作;区域 5 是代码编辑区;区域 6 是控制台输出界面。

调试工具栏中有若干按钮,功能如下。

(1)"新建文件"按钮,单击该按钮(或使用快捷键 Ctrl+N),新建一个空文件。

(2)"打开工程"按钮,单击该按钮(或使用快捷键 Ctrl+K、Ctrl+O),打开本地已有的

第1章 实验环境

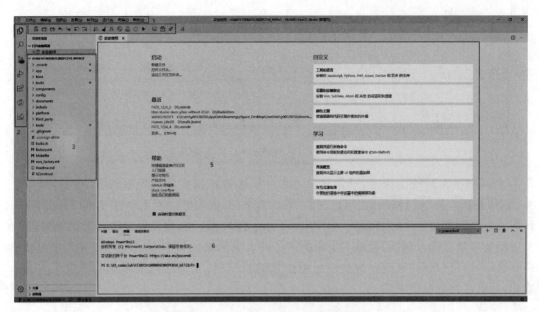

图1.3 LiteOS Studio 主界面构成区域

工程。

(3)"新建工程"按钮,单击该按钮,进入新建工程向导界面,可远程下载一个 LiteOS 工程的 SDK。

(4)"撤销"/"恢复"按钮,单击相应按钮(或使用快捷键 Ctrl＋Z/Ctrl＋Y),实现撤销/恢复上一步操作。

(5)"前进"/"后退"按钮,单击相应按钮(或使用快捷键 Alt＋←/Alt＋→),跳转到浏览历史中上一个/下一个页面。

(6)"编译"按钮,单击该按钮(或使用功能键 F7),对当前打开工程进行编译。

(7)"清理编译"按钮,单击该按钮(或使用功能键 F6),删除上一次编译生成的文件。

(8)"重新编译"按钮,单击该按钮(或使用功能键 Alt＋F7),删除上一次编译生成的文件,再次执行编译。

(9)"停止编译"按钮,单击该按钮(或使用功能键 Ctrl＋Shift＋F7),停止正在进行的编译。

(10)"烧录"按钮,单击该按钮(或使用功能键 F8),将程序烧录至目标板。

(11)"重启目标板"按钮,单击该按钮(或使用功能键 Ctrl＋Shift＋F9),重启开发板。

(12)"开始调试"按钮,单击该按钮(或使用功能键 F5),启动调试。

(13)"串口终端"按钮,单击该按钮,打开串口终端界面。

(14)"调测工具"按钮,单击该按钮,打开调测工具界面。

(15)"工程配置"按钮,单击该按钮(或使用功能键 F4),打开工程配置界面。

串口终端界面用于和设备进行通信,打开串口终端界面如图1.4所示。

串口终端界面从上到下分为4块区域。1号区域用于串口的设置和开关,可以设置端口等操作。需要在"端口"下拉列表中选择与目标板连接的实际串口号。"波特率"默认为115200,可根据实际情况修改。"校验位"默认为 None,可根据实际情况修改。"数据位"默

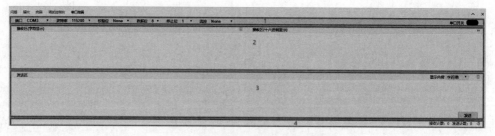

图 1.4 LiteOS Studio 串口终端界面

认为 8,可根据实际情况修改。"停止位"默认为 1,可根据实际情况修改。"流控"默认为 None,可根据实际情况修改。"串口开关"默认为关闭,使用时需要手动打开串口开关;2 号区域为串口数据接收区,左边显示字符串,右边显示十六进制,暂未设定接收数据量限制;3 号区域为串口数据发送区,通过右上角的下拉列表来切换数据内容由字符串显示还是十六进制显示,通过"发送"按钮将数据发送给连接的目标板;4 号区域展示接收和发送了多少数据,并可以将计数值清零。

1.2.2 STM32 工程工具

STM32 工程使用 makefile 文件进行构建管理,需要安装 make.exe 构建工具和 GNU Arm Embedded Toolchain 编译交叉工具链,并使用 JLink 仿真器。

安装 GNU Make 工具,可以通过执行 LiteOS 代码库 https://liteos.gitee.io/liteos_studio 提供的 x_pack_windows_build_tools_download 自动下载程序来进行下载,默认下载到 C:/Users/< UserName >/.huawei-liteos-studio/tools/build/bin 目录,如图 1.5 所示。

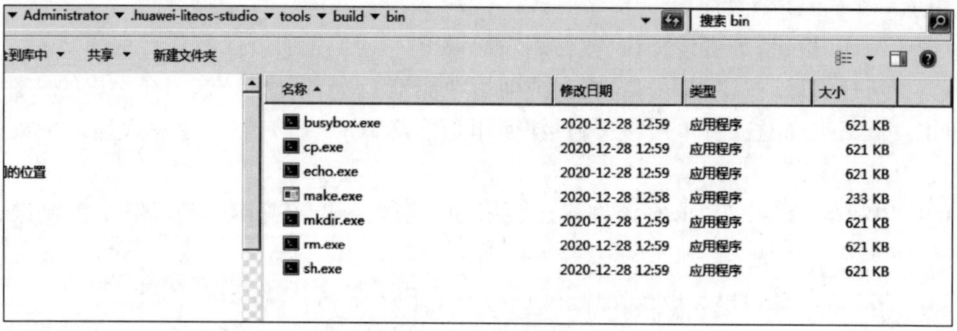

图 1.5 GNU Make 工具

该下载程序为 bat 批处理文件,功能为下载工具 xpack-windows-build-tools-2.12.2-win32-x64。不过,通过该下载程序下载 GNU Make,需要先安装 git for windows 工具,并加入操作系统环境变量。也可以直接在 Github 下载该工具并保存在前述默认目录下,该网址为 https://github.com/xpack-dev-tools/windows-build-tools-xpack/releases/download/v4.2.1-2/xpack-windows-build-tools-4.2.1-2-win32-x64.zip。解压后的 bin 目录下的文件含有 make.exe 文件。需要把该目录添加到操作系统的 PATH 环境变量中。

使用 ARM GCC 编译器将 LiteOS 工程编译为 ARM 架构的可执行程序,因此需要安装编译器软件。同前，可以通过执行 LiteOS 代码库提供的 GNU Arm Embedded Toolchain 自动下载程序来进行下载，默认下载到 C:/Users/< UserName >/.huawei-liteos-studio/tools/arm-none-eabi 目录。该自动下载程序下载的是 gcc-arm-none-eabi-9-2019-q4-major-win32，本书例子中的编译器版本为 gcc-arm-none-eabi-10-2020-q4-major-win32，默认安装在 Program Files（x86）/GNU Arm Embedded Toolchain/10 2020-q4-major 目录下，如图 1.6 所示。

图 1.6 gcc-arm-none-eabi 工具链

在 Linux 命令行工具和 shell 中，大多数命令默认以空格作为值与值之间的分隔符，而不是作为文件名的一部分。建议更换一下编译器的安装目录，避免在目录名称和文件名中使用空格。比如安装在 D:/armgcc 目录下，并把该目录 D:/armgcc/bin 添加到 PATH 环境变量中。

开发板如果使用 JLink 仿真器，则根据情况安装 JLink 软件。访问 https://www.segger.com/downloads/jlink/，选择 J-Link Software and Documentation Pack 后再选择 J-Link Software and Documentation pack for Windows 下载 JLink 软件，并按安装向导完成最新版 JLink 的安装。将 JLink.exe 所在目录加入操作系统 PATH 环境变量。

可以为 LiteOS Studio 安装中文语言包扩展 vscode-language-pack-zh-hans。安装过程是通过执行 LiteOS 代码库提供的扩展自动下载程序下载中文语言包扩展，默认下载到 C:/Users/< UserName >/.huawei-liteos-studio/extensions/extension-collections 目录。也可自行安装中文语言包扩展，需注意不同版本可能不兼容。低版本无法兼容高版本的中文插件包。

1.2.3 Hi3861V100 WiFi IoT 工程工具

Hi3861 处理器芯片为 RISC-V 架构，和 STM32 芯片不同，WiFi IoT SDK 软件使用 Scons 进行构建管理，因此需要安装 Python 和 Scons 库。WiFi IoT SDK 使用 riscv32-unknown-elf 编译器进行编译，使用 JLink 仿真器进行调测。

可以从官网 https://www.python.org/downloads/release/python-376/ 或者国内镜像网站如清华大学镜像下载 Python 3.7.6，按照安装向导完成 Python 3.7.6 的安装。安装 Python 时，勾选 Add Python ×.× to PATH 复选框，将安装 Python 的根目录，以及安装

根目录下的 Scripts 目录,加入到操作系统 PATH 环境变量。如不勾选 Add Python ×.× to PATH 复选框,需手动添加到操作系统 PATH 环境变量。

安装 Python 第三方库 Scons 时,使用 Pypi 镜像可以提升安装 Python 第三方库的速度。推荐使用华为云开源镜像,但需要修改 pip 文件设置镜像。对于 Windows 操作系统,在 C:/Users/<UserName>/pip 目录下添加 pip.ini 文件。如果不存在 pip 目录和 pip.ini 文件,则需要自己创建,然后编辑 pip.ini 文件,其内容如下。

```
[global]
index-url = https://repo.huaweicloud.com/repository/pypi/simple
trusted-host = repo.huaweicloud.com
timeout = 120
```

最后打开 Python 编辑器命令行窗口,执行如下命令安装 Python 第三方库。

```
pip install pycryptodome
pip install ecdsa
pip install pywin32
pip install scons
```

如果需要执行编译功能,则应安装相应的 gcc 交叉编译器,即 riscv32-unknown-elf 编译器。该编译器以开源的方式托管在 GitHub 上,网址为 https://github.com/riscv/riscv-gnu-toolchain。如果不需要源码,在 Windows 系统中直接下载 Windows 系统的安装包 xPack GNU RISC-V Embedded GCC。64 位系统的安装包下载网址为 https://github.com/xpack-dev-tools/riscv-none-embed-gcc-xpack/releases/download/v10.1.0-1.1/xpack-riscv-none-embed-gcc-10.1.0-1.1-win32-x64.zip。其他版本可以在 https://github.com/xpack-dev-tools/riscv-none-embed-gcc-xpack/releases/ 找到。

使用 JLink 仿真器,仍需安装 JLink 软件,请参考 1.2.2 节的内容。

1.2.4 DevEco Device Tool

DevEco Device Tool 是面向智能设备开发者提供的一站式集成开发环境,支持按需定制 HarmonyOS 的组件,支持编辑代码、编译、烧录和调试等功能,支持 C/C++ 语言,并以插件的形式部署在 Visual Studio Code 上。

DevEco Device Tool 可以在 Windows 和 Ubuntu 系统中应用,具有以下特点。

(1) 支持代码查找、代码高亮、代码自动补齐、代码输入提示和代码检查等功能,开发者可以轻松、高效地编码。

(2) 支持多种类型开发板,包括基于华为海思芯片的 Hi3516DV300、Hi3518EV300、Hi3861V100 和 BearPi-HM Nano 等开发板,以及三方厂商的 Imx6ull、Rtl8720、Xr872 和 Neptune 等开发板。

(3) 支持单步调试能力和查看内存、变量、调用栈、寄存器和汇编等调试信息。

(4) 支持 HDF 驱动管理,可在源码中快速创建驱动模板,并自动生成相关接口和依赖。

在 Windows 10 64 位系统中安装 DevEco Device Tool 的过程非常简单。在 Ubuntu 系统中安装 DevEco Device Tool 的过程类似,具体过程请参考官方网站 https://device.

harmonyos.com/cn/docs/documentation/guide/install_windows-0000001050164976。

DevEco Device Tool 当前最新版本是 devicetool-windows-tool-3.0.0.800，本书使用的版本是 devicetool-windows-tool-3.0.0.200。DevEco Device Tool 以插件方式提供，基于 Visual Studio Code 进行扩展，同时需要 Python、Node.js 和 hpm 工具。首先准备好相关软件，依次安装 Visual Studio Code、Python、Node.js 和 hpm，最后安装 DevEco Device Tool 插件。如果未提前准备好这些软件，新版本 DevEco Device Tool 支持一体化安装，安装前安装程序先检测操作系统是否安装 Visual Studio Code、Python、Node.js 和 hpm 的适配版本，会提示是否安装，如图 1.7 所示。

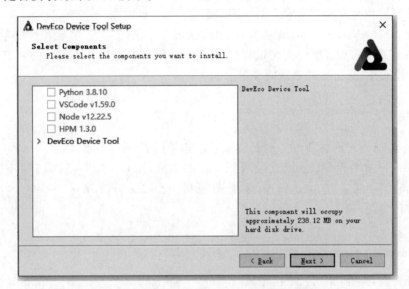

图 1.7　安装 DevEco Device Tool 的构件选择

如果未安装，则需勾选这些构件。建议在安装 DevEco Device Tool 之前安装好这些软件。

Visual Studio Code 是微软开发的代码编辑工具，需要 V1.53 及以上 64 位版本。Python 是编译构建工具，版本要求 3.8～3.9，x64 位版。Node.js 提供 npm 环境，需要 v14.17.5 及以上 64 位版本。hpm 包管理工具，需要最新版 v1.3.0 及以上。

如需检查是否已安装 Visual Studio Code，可以在"命令提示符"界面输入"code -version"命令，检查版本号。若可正常返回版本号，说明操作系统的 PATH 环境变量设置好了，不需要重新安装。建议下载最新版本的 Visual Studio Code 并安装。进行安装过程中，勾选"添加到 PATH（重启后生效）"选项。安装完成后检查一下，本书使用 1.61.2 版本。按组合键 Win+R 并输入 cmd，进入"命令提示符"界面进行检查。检查的结果如下。

```
D:\> code -- version
1.61.2
6cba118ac49a1b88332f312a8f67186f7f3c1643
x64
```

安装 Python 时，勾选 Add Python ×× to PATH 选项，然后进行安装。如果安装 DevEco Device Tool 2.1 Release 版本，Python 版本只能为 3.8.×，不能安装最新的 Python

3.9.×及以上版本。如果安装DevEco Device Tool V3.0 Beta1及以上版本,Python版本只能为3.8.×或3.9.×。本书实验使用的Python版本为3.8.9。按组合键Win+R输入"cmd"进入"命令提示符"界面,在命令行中输入"python",可以进入Python环境查看版本号,然后输入"quit()"退出,如下。

```
D:\> python
Python 3.8.9 (tags/v3.8.9: a743f81, Apr 6 2021, 14:02:34) [MSC v.1928 64 bit (AMD64)]
on win32 Type "help", "copyright", "credits" or "license" for more information.
>>> quit()
```

如果已安装Node.js,打开命令行工具,输入"node -v"命令,检查版本号。如未安装,则在网站https://nodejs.org/zh-cn/download/下载安装。安装向导中,全部按照默认设置单击Next按钮,直至完成安装。安装过程中,Node.js会自动在操作系统PATH环境变量中配置可执行文件node.exe的目录路径。

```
D:\> node -v
v12.22.5
```

安装hpm前需先确保Node.js安装成功。如果已安装hpm,可以执行npm update -g @ohos/hpm-cli命令升级hpm至最新版本。在安装hpm前,网络连接状态需能访问Internet,同时需要先设置npm代理,才能安装hpm。按组合键Win+R输入"cmd"进入"命令提示符"界面,输入如下命令。

```
npm config set registry https://repo.huaweicloud.com/repository/npm/
npm install -g @ohos/hpm-cli
```

完成安装后,打开命令行工具,输入"hpm -V"检查hpm安装结果。hpm -V中的V必须大写。

```
D:\> hpm -V
1.4.0
```

建议下载最新版本的DevEco Device Tool,目前版本是3.0 beta2版,下载网址为https://device.harmonyos.com/cn/develop/ide#download_beta。

安装完成后,启动Visual Studio Code,会自动安装DevEco Device Tool依赖的C/C++、CodeLLDB插件。等待安装完成后,单击Visual Studio Code左侧的"扩展"按钮,检查INSTALLED页中,是否已有C/C++、CodeLLDB和DevEco Device Tool插件。如果C/C++和CodeLLDB插件在线安装失败,可以采用离线方式进行安装。

离线安装前需要下载C/C++和CodeLLDB插件。在网页https://github.com/microsoft/vscode-cpptools/releases下载插件安装文件cpptools-win32.vsix,此为Windows版本的C/C++。在网页https://github.com/vadimcn/vscode-lldb/releases下载插件安装文件codelldb-x86_64-windows.vsix,此为Windows版本的CodeLLDB。下载完后安装这两个插件,如图1.8所示。

运行Visual Studio Code,选择Views and More Actions,然后选择install from VSIX,分别安装C/C++和CodeLLDB插件。如为中文版Visual Studio Code,则选择"视图和更多操作",然后选择"VSIX安装"。

图 1.8 离线安装插件

1.3 LiteOS Studio 实验环境

1.3.1 LiteOS Studio 工程配置

运行 LiteOS Studio 后,通过单击"新建工程"按钮打开"新建工程"界面,如图 1.9 所示。

图 1.9 LiteOS Studio"新建工程"界面

新建工程需使用工具 git 进行 LiteOS SDK 下载,需要预置 git for Windows 工具,可访问 git for Windows 官网自行下载 git 工具。如果无法联网或 git 失败,可以把在 LiteOS 网站下载下来的 LiteOS SDK 复制至相应目录。"新建工程向导"界面中,需要设置如下 4 项内容。

(1) 工程名称。在"工程名称"框中填写自定义的工程名称,作为 SDK 工程的根目录名。

(2) 工程目录。在"工程目录"框中填写 SDK 工程的本地存储路径,建议路径名中不要包含中文、空格、特殊字符等字符。

(3) SDK 版本。在"SDK 版本"框中选择 LiteOS 的不同版本,从而在下面的"开发板信息表"中显示不同版本支持的开发板。

(4) 开发板信息表。页面下半部分区域的表格面板,能够展示所选版本支持的开发板信息,包括开发厂商、开发板名称、对应设备名称与内核名称。

通过工具栏中的"工程配置"按钮(或使用功能键 F4)打开"工程配置"界面,如图 1.10 所示。

图 1.10　LiteOS Studio 工程目标板配置

单击"工程配置"页面左侧的"目标板"选项进入"目标板配置"界面,选择目标板信息面板上的匹配开发的行后,单击"确认"按钮保存设置,即指定了当前工程的开发板,后台将根据开发板设置默认的编译、烧录等配置信息。"目标板配置"界面支持用户自行添加目标板信息,单击信息面板上的"+"按钮,即可增添一行空行,其中,"厂商""目标板名称""设备名称"和"内核名称"四栏必须填写,填写完成后,按回车键。用鼠标选中刚添加的一行,单击"确认"按钮保存,即可使用新增的目标板信息进行后续配置。对于自行添加的信息行,鼠标移至该行上时，操作栏将出现"-"按钮，单击该按钮即可删除该行。同时，在自行添加的行上双击，即可重新对该行进行编辑。新增目标板需要在编译、烧录、调试等方面满足 LiteOS Studio 所支持的工具与架构。当前版本 LiteOS Studio 仅支持 ARM/RISCV32 的编译方式、JLink/HiBurner/OpenOCD 的烧录方式和 JLink 的调试方式。同时，对应的工程源码也应完成在 Windows 上的适配，如所使用的工具、架构或编译、烧录、调试等流程。如果所使用的命令与预置的几款开发板有明显差异，则不支持自行添加开发板。

单击"工程配置"页面左侧的"组件配置"选项进入"组件配置"界面，首次启动仅展示本地已有的组件。单击左侧"组件列表"中的某项，在右侧"组件属性"栏通过勾选为组件使能，或输入具体的参数值，单击"确认"按钮保存后，LiteOS Studio 将在后台打开组件对应的宏开关，将使能的组件与更新后的属性值加入编译。可以使用快捷键 Ctrl+F 调出组件搜索框，填入关键字进行搜索，若匹配成功，右侧"组件属性"区域将显示搜索结果，此时能够单击"向上""向下"箭头切换搜索结果，搜索框提供了区分大小写、全字匹配和使用正则表达式三种模式。

单击"工程配置"页面左侧的"编译器"选项进入"编译器"界面。目前"编译器类型"框有 arm-none-eabi、riscv32-unknown-elf 两种编译器，分别适用于 ARM 架构和 RISC-V 架构的开发板，默认已经配好，如无额外需求不需要配置。"编译器目录"是用户所使用的编译器所在目录。LiteOS Studio 不提供编译器预置，需要自行安装。在 Windows 系统下需要安装相关文件，安装过程请参考 1.2.2 节的 ARM GCC 编译器 arm-none-eabi 安装和 1.2.3 节的 riscv gcc 编译器安装。对于 ARM 架构的开发板，包括 Qemu realview-pbx-a9 仿真器在内，"编译器目录"框里填写 arm-none-eabi-gcc.exe 文件所在路径，如本书的 D:/armgcc/bin。对于 RISC-V 架构的开发板，"编译器目录"框填写 riscv32-unknown-elf-gcc.exe 文件所在路径。

配置项"Make 构建器"仅在开发板支持 makefile 脚本与 arm-none-eabi 编译方式时出现，LiteOS Studio 不提供构建器预置，需要自行安装。可参考 1.2.2 节安装 make 构建软件，"Make 构建器"框中，路径填写文件 make.exe 所在目录，如本书中为 C:/Users/Administrator/.huawei-liteos-studio/tools/build/bin。

配置项"Makefile 脚本和 SConstruct 脚本"框中，是按照目标板的编译架构不同，所显示的编译脚本路径配置项，目前 LiteOS Studio 支持 makefile 和 Scons 两种编译脚本，支持自动配置与手动配置。编译脚本路径建议不要包含中文、空格和特殊字符，避免编译失败。

手动配置脚本有两种方式可以配置。第一种方式是单击"目录"按钮浏览目录自行配置；第二种方式是右击 makefile 或 SConstruct 文件，在弹出的快捷菜单中选择"设置为 MakeFile/SConstruct 文件"完成设置，设置完成后会自动在编译脚本路径配置框中填入脚本路径。如图 1.11 所示为配置 SConstruct 脚本的过程。如是 ARM 架构，此处需填写"工程目录"为/LiteOS-master/Makefile 目录。

自动配置脚本更为简单，单击"自动搜索脚本"按钮自动匹配当前开发板对应的 Makefile/SConstruct 脚本，如果自动搜索结果为空，或使用该搜索结果导致编译报错，可能由于工程脚本名称、路径等发生改变，可改为手动配置方式设置脚本路径。

Make/SCons 参数是执行编译时可以自行添加的参数，例如"-j 32"等，该参数自行配置。

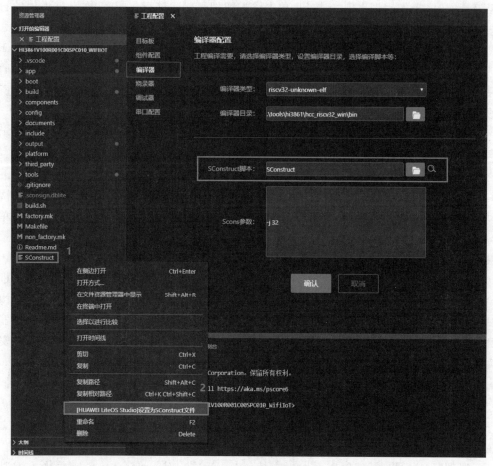

图 1.11 配置脚本

填写完成后,单击"确认"按钮保存用户配置,此时可以通过单击工具栏中的"编译"按钮,对当前工程进行编译。如果使用 Qemu realview -pbx-a9 仿真器,需要正确配置 QemuDebug 选项。

单击"工程配置"页面左侧的"烧录器"选项进入"烧录器配置"界面。"烧录方式"下拉菜单目前支持 JLink、HiBurner 和 OpenOCD 三种烧录器,以及 Simulator 仿真器。默认已经配好,如无额外需求不需要用户配置。"烧录器目录"框是所使用的烧录器所在目录。

LiteOS Studio 仅预置 HiBurner 烧录器,默认安装在 C:/users/用户名/.huawei-liteos-studio/tools 目录下,不需要自行配置此目录。JLink 烧录器目录需要将 JLink.exe 所在目录加入操作系统的 PATH 环境变量,"烧录器目录"文本框填写 JLink.exe 所在路径。OpenOCD 烧录器需要在"烧录器目录"文本框填写 openocd.exe 所在路径。"Simulator 仿真器目录"文本框填写 qemu-system-arm.exe 所在路径。烧录文件是编译生成的文件,目前支持 bin、fwpkg 和 hex 三种格式。执行编译后,后台将保存生成的烧录文件(bin、fwpkg、hex 后缀的文件),并填入"烧录文件"下拉菜单中供用户选择,也可以在 bin、fwpkg 或 hex 文件上右击,在弹出的快捷菜单上选择"设置为烧录文件"选项,或单击"目录"按钮浏览目录自行配置。

单击"工程配置"页面左侧的"串口配置"选项进入"串口配置"界面。"串口配置"界面包括端口、端口状态、波特率、数据位、停止位、奇偶和流控等设置,大多选项默认配好,如有额外需求,可自行配置。"端口"文本框填写开发板使用的端口号,一般使用 HiBurn 烧录的开发板在烧录前需要手动配置,可通过操作系统的"设备管理器"界面来判断当前使用的端口号。填写完成后,单击"确认"按钮保存配置,此时可以通过单击"烧录"按钮对当前工程进行烧录。

LiteOS Studio 集成了 AI 模型代码生成功能,通过 msmicro 工具,将压缩包中的 LiteOS_Studio/tools/mindspore_micro/msmicro.rar 解压到本地。AI 模型生成的原理是将 MindSpore 训练的模型或第三方模型转换为 ms 模型,并将 ms 模型解析为算子,生成 .c 文件或指令集优化的汇编代码。然后通过交叉编译器,编译支持不同平台的可执行文件到 IoT 设备部署。tensorflow_lite 模型文件训练与生成方式可参考 tensorflow 官方社区文档操作。tensorflow_lite 应用实例网址参考 https://www.tensorflow.org/lite/examples?hl=zh-cn。

"AI 模型代码生成向导"界面可配置的参数包括 Msmicro 目录、框架类型、AI 模型文件、量化类型和配置文件,需要自行配置。"Msmicro 目录"文本框是用户自行下载并解压 msmicro 工具后,填入 msmicro.exe 所在目录。"框架类型"包括 TF、CAFFE、ONNX、MS 和 TFLITE 五种。AI 模型文件需要自行获取或自主生成,在"AI 模型文件"文本框中填入模型文件所在路径。"量化类型"包括 AwareTraining、PostTraining 和 WeightQuant 三种。当使用 Mnist.tflite,并需要进行训练后量化时,选择 PostTraining。"配置项输入框"仅在量化类型选择了 PostTraining 时出现,需要填入相应的配置文件,当使用 Mnist.tflite 并选择了 PostTraining 量化类型时,需要填入 config.mnist 所在路径,config.mnist 中需要填入本地校准集的绝对路径。配置完成后,单击"确定"按钮,即可开始代码生成,并自动将生成的文件放入相应的编译路径下。

1.3.2 STM32 工程环境

LiteOS Studio 支持 STM32F429IG、STM32L431RC、STM32F769NI、STM32F072RB、STM32F103ZE 和 STM32F407ZG 等 STM32 系列开发板,以及 Qemu 仿真的 realview-pbx-a9 开发板。

Windows 系统下,开发环境需要安装 make 构建工具、GNU Arm Embedded Toolchain 编译交叉工具链、JLink 仿真器、OpenOCD 烧录工具、USB 转串口驱动程序和 Qemu 仿真器等。

LiteOS Studio 新建工程的步骤如下。

(1) 在"工程名称"框中填入自定义的工程名。

(2) 在"工程目录"框中填入或选择工程存储路径,路径名中不要包含中文、空格和特殊字符等。

(3) 选择 SDK 版本号,当前 STM32 工程被维护在 https://gitee.com/,支持最新版本 master 分支。

(4) 在开发板信息表选择开发板所在行,如 realview-pbx-a9 开发板。

单击"确认"按钮,后台将下载并保存所选目标板的 SDK,等待下载完成后会在一个新窗口中自动打开新建的工程。如果无法在线下载,可以把源码复制到工程存储路径中。

打开工程后,单击工具栏上的"工程设置"按钮,打开"工程配置"界面。"目标板配置"框中选择目标板,如 realview-pbx-a9 开发板。LiteOS 开放可配置的组件与属性进行使能,而不是仅使用默认的配置,可以单击"工程配置"界面上的"组件配置"页面,在左侧的"选择组件"中单击想要使能或修改的组件,在右侧的组件属性栏中勾选需要使能的组件,或更改组件属性值,单击"确认"按钮保存。

编译配置的过程如下。

(1) 单击"工程配置"界面上的"编译器"按钮。

(2) "编译器类型"框选择 arm-none-eabi。

(3) 指定"编译器的安装目录",设置为 arm-none-eabi-gcc.exe 所在路径。

(4) 指定 make.exe 构建工具的安装目录,设置为 make.exe 所在路径。

(5) "Makefile 脚本路径"框已填入默认值。对于 STM32 工程,在工程根目录下的 makefile 文件上单击鼠标右键,设置为 makefile 文件。

(6) "Make 参数"框已填入默认值,根据系统的配置确定。如果计算机性能较差,填入"-j 2"即可。

(7) 配置好后单击"确认"按钮进行保存。

(8) 单击工具栏上的"编译"按钮开始编译,也可以单击"重新构建"按钮进行清理和重新编译。

如果没有配置错误,编译成功的话,在 Terminal 窗口中会给出提示"LiteOS build successfully!"并在该工程文件下创建了 out 目录,用于存放烧录文件,如 bin 和 elf 文件等。

直连烧录方式的过程如下。

(1) 单击"工程配置"界面上的"烧录器"选项。

(2) 对于目标板,"烧录方式"框选择 JLink。如果目标板是 STM23L431RC 和 STM32F769NI,建议刷成 JLink 进行调测。

(3) "烧录器目录"框已提供默认路径 C:/Program Files (x86)/SEGGER/JLink,如果 JLink 未安装在此目录,需要根据实际目录进行设置。

(4) "烧录文件"框的设置,如图 1.12 所示。

目标板是 Qemu 仿真 realview-pbx-a9 的烧录器配置如图 1.13 所示。

进入"烧录器配置"界面,"烧录方式"框选择 Simulator,"烧录器目录"框选择 qemu-system-arm.exe 所在目录,"烧录文件"框选择 out/realview-pbx-a9 目录下的 Huawei_LiteOS.bin 文件。

(5) "连接方式"框、"连接速率"框和"加载地址"框等保持默认,或根据开发板进行调整。配置好后单击"确认"按钮保存。

(6) 单击工具栏上的"烧录"按钮进行烧录,仿真启动后会运行 realview-pbx-a9 工程并进入 shell 交互界面。

如果需要重新编译工程,在编译前需要先退出 shell 交互界面,此时在交互界面所在的终端界面右上角单击"终止终端"按钮即可退出。

LiteOS Studio 调测可支持 STM32 开发板的图形化单步调试。调测设置过程如下。

第1章 实验环境

图1.12 LiteOS Studio 烧录器配置

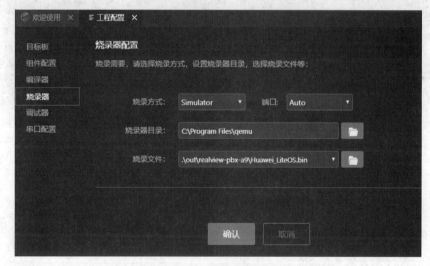

图1.13 LiteOS Studio 烧录器配置 Simulator

(1) 单击"工程配置"界面上的"调试器"选项进入"调试器配置"界面。

(2) "调试器"框根据实际情况选择 JLink、OpenOCD 或者 Simulatior。对于板载 ST-Link 仿真器的开发板,建议刷成 JLink 进行调测。

(3) "连接方式"框根据需求选择 SWD 或 JTAG,"连接速率"框可以默认或者自行指定,如果是 Simulatior 方式,则没有这一步。

(4) "调试器目录"框根据实际情况设置为 JLink 或者 OpenOCD 调试器的安装目录,如果与实际安装目录不一致,调试可能失败。选择 Simulator 后,调试器目录选择 qemu-system-arm.exe 文件所在目录,本书中即 D:/qemu/。

(5)"GDB 目录"框设置编译器目录,即 armgcc 的 bin 目录。

(6)完成编译后就会生成可执行文件 elf,可以手动设置,也可以选择自动匹配的可执行文件。

(7)"调试配置"框中,根据需要,选择复位调试或附加调试。复位调试会自动重启开发板,并停止在 main()函数。附加调试不重启开发板,附加到当前运行代码行。配置好后单击"确认"按钮保存。

(8)单击"调试并运行"按钮(或按快捷键 Ctrl+Shift+D),根据调测器配置,选择调试配置为 JLINK Debug 或 OpenOCD Debug 或 Qemu Debug,单击绿色三角按钮,开始调试。调试界面如图 1.14 所示。

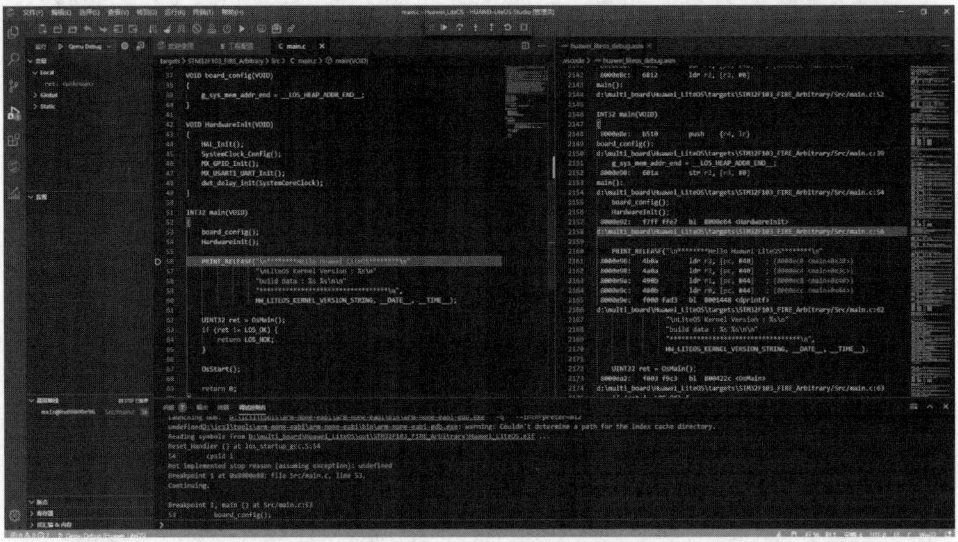

图 1.14 LiteOS Studio 调试界面

调试界面分为变量区、调用堆栈区、寄存器区、反汇编 & 内存区、输出区和调试控制台等区域。

- 变量区展示局部变量、全局变量、静态变量。
- 监视区监视指定的表达式。
- 调用堆栈区展示当前运行状态和暂停状态的任务调用堆栈。
- 断点区展示设置的断点。
- 寄存器区查看各个寄存器的数值,支持复制数值操作。
- 反汇编 & 内存区支持对函数进行反汇编,支持查看内存操作。
- 输出区展示 GDB 客户端的输出日志。
- 调试控制台展示 GDB Server 的输出日志。

LiteOS Studio 中添加断点时,可在代码行号处单击添加为断点。也可以通过右击菜单添加断点、添加条件断点和添加记录点。和其他编译器类似,可添加变量或表达式,可以添加监视点。在"调试"面板的"寄存器"视图中,可查看开发板 CPU 中寄存器的数值。通过多线程感知调测技术，调测时可以展示 LiteOS 中运行态和暂停态的任务线程的调用堆栈。在"调试"面板的"反汇编 & 内存"视图中，单击"查看内存"按钮，在弹窗中输入内存起始地址及长度，可以展示开发板的内存信息。

还可以使用 LiteOS Studio 通过 JLink 服务实现远程烧录和远程调试。

1.3.3 LiteOS shell

LiteOS 提供 shell 命令行界面，它能够以命令行交互的方式访问操作系统的功能或服务。它接收并解析用户输入的命令，并处理操作系统的输出结果。作为在线调试工具，可以通过串口工具输入输出，支持常用的基本调试功能。同时可以新增定制的命令，新增命令需重新编译烧录后才能执行。

LiteOS 的 shell 中的系统命令在使用前，需要先通过 make menuconfig 使能 shell。LiteOS 提供了 help、task 等多个系统命令。如 help 命令功能是用于显示当前操作系统内所有的 shell 命令。

task 命令用于查询系统任务信息，命令格式为 task [ID]。参数为任务 ID，取值范围为以十进制表示或十六进制表示的数[0,0xFFFFFFFF]。参数省略时，默认打印全部运行任务信息。task 后加 ID，当 ID 参数在[0,64]范围内时，返回指定 ID 任务的任务名、任务 ID、任务的调用栈信息（最大支持 15 层调用栈），其他取值时返回参数错误的提示。如果指定 ID 对应的任务未创建，则提示。如果在 task 命令中，发现任务是 Invalid 状态，请确保 pthread_create()创建函数时有进行如下操作之一，否则资源无法正常回收。选择的是阻塞模式应该调用 pthread_join()函数。选择的是非阻塞模式应该调用 pthread_detach()函数。如果不想调用前面两个接口，就需要设置 pthread_attr_t 状态为 PTHREAD_STATE_DETACHED，将 attr 参数传入 pthread_create，此设置和调用 pthread_detach()函数一样，都是非阻塞模式。task 命令输出项如表 1.1 所示。

表 1.1 task 命令输出项

输 出 项	说 明
Name	任务名
TID	任务 ID
Priority	任务的优先级
Status	任务当前的状态
StackSize	任务栈大小
WaterLine	该任务栈已经被使用的内存大小
StackPoint	任务栈指针，表示栈的起始地址
TopOfStack	栈顶的地址
EventMask	当前任务的事件掩码，没有使用事件，则默认任务事件掩码为 0（如果任务中使用多个事件，则显示的是最近使用的事件掩码）
SemID	当前任务拥有的信号量 ID，没有使用信号量，则默认为 0xFFFF（如果任务中使用了多个信号量，则显示的是最近使用的信号量 ID）
CPUUSE	系统启动以来的 CPU 占用率
CPUUSE10s	系统最近 10s 的 CPU 占用率

续表

输　出　项	说　　明
CPUUSE1s	系统最近 1s 的 CPU 占用率
MEMUSE	截止到当前时间,任务所申请的内存大小,以 B 为单位显示。MEMUSE 仅针对系统内存池进行统计,不包括中断中处理的内存和任务启动之前的内存。任务申请内存,MEMUSE 会增加,任务释放内存,MEMUSE 会减小,所以 MEMUSE 会有正值和负值的情况
(1) MEMUSE 为 0	说明该任务没有申请内存,或者申请的内存和释放的内存相同
(2) MEMUSE 为正值	说明该任务中有内存未释放
(3) MEMUSE 为负值	说明该任务释放的内存大于申请的内存

任务状态如下。

(1) Ready 表示任务处于就绪状态。Pend 表示任务处于阻塞状态。

(2) PendTime 表示阻塞的任务处于等待超时状态。

(3) Suspend 表示任务处于挂起状态。

(4) Running 表示该任务正在运行。

(5) Delay 表示任务处于延时等待状态。

(6) SuspendTime 表示挂起的任务处于等待超时状态。

(7) Invalid 表示非上述任务状态。

1.4　DevEco Device Tool 实验环境

进行设备开发时,需要编辑源代码、编译代码、烧写和调试系统。使用 DevEco Device Tool 编辑代码,可在 Ubuntu 系统或者 Windows 系统下进行。使用 DevEco Device Tool 编译代码,通常在 Ubuntu 系统进行。源码的编译,也可在 Docker 容器进行,Windows 系统和 Ubuntu 系统适用。对于 Hi3861,支持使用 DevEco Device Tool 在 Windows 下编译。烧写系统映像,通常由 DevEco Device Tool 在 Windows 下进行。

1.4.1　Windows 系统下编译

这里的硬件使用 HiSpark 的 WiFi IoT 套件,该套件的主板为 Hi3861 核心板。

Hi3861 支持 Windows 系统的编译,需要手动下载工具链。所需工具包括 ninja、gn、gcc_riscv32 和 tool_msys 等。下载地址如下。

(1) ninja：https://repo.huaweicloud.com/harmonyos/compiler/ninja/1.9.0/windows/ninja-win.zip。

(2) gn：https://repo.huaweicloud.com/harmonyos/compiler/gn/1744/windows/gn-windows-amd64.zip。

(3) gcc_riscv32：http://www.hihope.org/download/download.aspx? mtt=34。

(4) tool_msys：https://sourceforge.net/projects/mingw/。

工具 ninja 和 gn 为单个文件,gcc_riscv32 则有一组文件,tool_msys 需要在线安装。注意下载时要记住这些工具的版本号。

这些工具下载完后,需要在 DevEco Device Tool 中进行设置。首先进入"导入工程"界面,如图 1.15 所示。

图 1.15　DevEco Device Tool 导入工程

将下载的 OpenHarmony SDK 解压到一个目录,用 Visual Studio Code 打开该目录。在 DevEco Home"工程"栏里,单击"导入工程"按钮。在"导入工程"框中,找到 OpenHarmony SDK 目录,并单击"导入"按钮。然后,"MCU"选择 Hi3861,"编译框架"选择 hb。

然后将以上四个工具添加到里面,添加过程如图 1.16 所示。

图 1.16　DevEco Device Tool 导入工具

在添加工具过程中，注意工具的名称和位置。工具的"组件名称"列表中，ninja 设置的路径为文件 ninja.exe 所在目录，gn 设置的路径为文件 gn.exe 所在目录，gcc_riscv32 设置的路径为 hcc_riscv32_win\bin 目录，tool_msys 设置的路径为 MinGW\msys\1.0\bin 目录。

工具添加完成后，单击"工程"及"配置工程"，选择 hi3861，在"执行配置"界面勾选"显示高级选项"，在 platform_packages 中，将四个编译工具添加进去，同时添加自带的 tool_scons，如图 1.17 所示。添加完成后单击"保存"按钮。

图 1.17　DevEco Device Tool 配置工程

配置完成后打开工程，单击 build 命令进行编译，在不添加文件的前提下，编译文件约有 300 个。

1.4.2　Docker 环境下编译

OpenHarmony 提供了两种 Docker 环境，以简化开发环境准备工作。一种 Docker 环境是独立 Docker 环境，适用于直接在 Ubuntu、Windows 操作系统平台进行 OpenHarmony 编译的场景。另一种是基于 HPM 的 Docker 环境，适用于使用 HPM 工具编译 HarmonyOS 的 Release 版本的场景。

这里的硬件使用 HiSpark 的 ipcamera 套件，该套件的主板为 Hi3516DV300 芯片。这里仅介绍独立 Docker 环境下，OpenHarmony 小型系统的编译。

首先下载 Docker，在官方网站下载的最新版 Windows 系统下 Docker 需要较新的

Windows 10 操作系统。可在镜像站点下载老版本的 Docker，如阿里云镜像站点 http://mirrors.aliyun.com/docker-toolbox/windows/docker-for-windows/beta/InstallDocker.msi。安装老版本 Docker 完成后，如提示是否更新，选择不更新。Docker 需要 Hyper-V 支持，安装完成后，提示重启并开启 Hyper-V。

在"开始"菜单里打开 Docker，查看版本等信息，如下。

```
C:\Users\fg> docker version
Client:
 Version:      17.03.1-ce-rc1
 API version:  1.27
 Go version:   go1.7.5
 Git commit:   3476dbf
 Built:        Wed Mar 15 20:33:22 2017
 OS/Arch:      windows/amd64

Server:
 Version:      17.03.1-ce-rc1
 API version:  1.27 (minimum version 1.12)
 Go version:   go1.7.5
 Git commit:   3476dbf
 Built:        Wed Mar 15 20:28:18 2017
 OS/Arch:      linux/amd64
 Experimental: true
```

拉取一个 hello-world 镜像，测试是否安装成功。

```
C:\Users\fg> docker run hello-world
Unable to find image 'hello-world:latest' locally
latest: Pulling from library/hello-world
2db29710123e: Pull complete
Digest: sha256:cc15c5b292d8525effc0f89cb299f1804f3a725c8d05e158653a563f15e4f685
Status: Downloaded newer image for hello-world:latest

Hello from Docker!
This message shows that your installation appears to be working correctly.
...
```

从 HuaweiCloud SWR 网站获取 OpenHarmony 的 Docker 镜像，如下。Docker 镜像文件约 1GB 大小。

```
D:\> docker pull swr.cn-south-1.myhuaweicloud.com/openharmony-docker/openharmony-docker:0.0.5

0.0.5: Pulling from openharmony-docker/openharmony-docker
83ee3a23efb7: Pull complete
db98fc6f11f0: Pull complete
f611acd52c6c: Pull complete
fa1e8123fd11: Pull complete
8b29c29307c0: Downloading [ ===>              ]  82.5 MB/1.048 GB
```

也可先获取 Dockerfile 脚本文件，再进入 Dockerfile 代码目录路径，通过执行 Docker

镜像构建命令构建本地 Docker 镜像。build 过程耗时较长，需要下载一系列 archieves，如下。

```
D:\> git clone https://gitee.com/openharmony/docs.git
#克隆到docs目录中
Cloning into 'docs'...
remote: Enumerating objects: 909, done.
remote: Counting objects: 100% (909/909), done.
remote: Compressing objects: 100% (665/665), done.
remote: Total 17408 (delta 519), reused 485 (delta 244), pack-reused 16499
Receiving objects: 100% (17408/17408), 211.42 MiB | 946.00 KiB/s, done.
Resolving deltas: 100% (10459/10459), done.
Updating files: 100% (4028/4028), done.

D:\> cd docs/docker
#注意最后有一个点
D:\docs\docker> docker build -t openharmony-docker:0.0.5 .
```

使用命令 docker image ls 查看已经获取的映像。

```
D:\> docker image ls
REPOSITORY TAG                IMAGE ID         CREATED         SIZE
hello-world latest            feb5d9fea6a5     2 months ago    13.3 kB
swr.cn-south-1.myhuaweicloud.com/openharmony-docker/openharmony-docker   0.0.5
c752ada0cfa3       5 months ago      2.7 GB
```

如果 OpenHarmony SDK 的源码放在 D:/OpenHarmony3.0 目录下，执行如下命令，从而进入 Docker 构建环境。

```
D:\> docker run -it -v D:\OpenHarmony3.0:/home/openharmony swr.cn-south-1.myhuaweicloud.com/openharmony-docker/openharmony-docker:0.0.5
root@e75db32dec2d:/home/openharmony#
```

编译脚本启动小型系统类设备的编译过程包括一些步骤，下面以 Hi3516 平台为例介绍。

(1) 通过"hb set"设置编译路径，命令如下。然后用键盘输入一个点"."选择当前路径。
root@e75db32dec2d:/home/openharmony# hb set

(2) 选择 ipcamera_hispark_taurus 并按回车键，如图 1.18 所示。

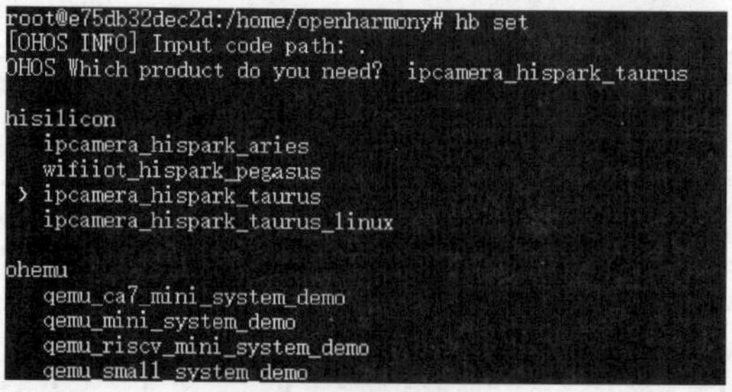

图 1.18　编译设置选择

(3) 通过"hb build -f"执行编译,命令如下。

root@e75db32dec2d:/home/openharmony# hb build -f

小型系统编译过程比轻量系统编译过程耗时,编译结果文件生成在 out/hispark_taurus/ipcamera_hispark_taurus 目录下。

(4) 使用 exit 命令退出 docker 环境。

1.4.3 烧录

编译成功后,在"配置工程"界面,选择 hi3861 页面,在 upload_port 框选择对应串口端号,在 upload_protocol 框选择 burn-serial,单击"保存"按钮。在保存成功后单击 upload 命令进行烧录,在"串口终端"出现"connecting, please reset device"字样后,复位核心板等待烧录完成。

烧录成功后,再次按下 Hi3861 核心板上的 RST 复位键,此时开发板的系统会运行起来。结果在 OLED 屏正中央,显示"Hello World",主板上的 LED 灯间隔 1s 交替闪烁。

1.4.4 调试

除了通过串口烧录,也可以使用 OpenOCD 板烧录,同时进行调试,此时需要硬件 OpenOCD 板,并将该板接入到 Hi3861 底板上。更多调试详情,请参考 https://device.harmonyos.com/cn/docs/documentation/guide/debug_overview-0000001050164998。

1.5 DevEco Studio 实验环境

DevEco Studio 基于 IntelliJ IDEA Community 开源版本,是面向华为终端全场景多设备的一站式集成开发环境(IDE),提供工程模板创建、开发、编译、调试和发布等 E2E 的 HarmonyOS 应用/服务开发。DevEco Studio 可以更高效地开发具备 HarmonyOS 分布式能力的应用/服务。

当前 DevEco Studio 的最新版本是 DevEco Studio 3.0 Beta3,可在华为开发者网站 https://developer.harmonyos.com/cn/develop/deveco-studio#download_beta 下载最新版本的软件,完成后,双击下载的 deveco-studio-3.0.0.800.exe,进入 DevEco Studio 安装向导,单击 Next 按钮,直至安装完成。

DevEco Studio 提供 SDK Manager 统一管理 SDK 及工具链,下载各种编程语言的 SDK 包时,SDK Manager 会自动下载该 SDK 包依赖的工具链。

第一次使用 DevEco Studio,需要下载 SDK 及对应工具链。后续使用中,可以更新部分组件,但更新可能带来软件版本不兼容的情况。因此不建议单独更新某组件,最好重新安装新版本的 DevEco Studio 及 SDK 包和工具链。

第2章

LiteOS实验

读书勤乃有，不勤腹中虚。

——韩愈

2.1 LiteOS 实验概览

1. 实验目的

本章实验的目的是掌握 LiteOS 操作系统中任务管理、内存管理、中断异常管理和任务间通信等基础模块的使用。

2. 实验设备

（1）硬件设备为一台主流配置的计算机。

（2）软件环境主要是 Windows 10 操作系统、LiteOS SDK 5.0、LiteOS Studio 1.45 和 Qemu 仿真器。详细的环境配置和安装过程请参考第 1 章。

3. 实验内容

包含 18 个实验，分别为：

（1）LiteOS 代码目录。

（2）创建任务实验。

（3）多核任务创建实验。

（4）调试任务实验。

（5）动态内存分配实验。

（6）静态内存分配实验。

（7）中断处理和错误处理实验。

（8）消息队列实验。

（9）事件实验。

（10）信号量实验。

（11）互斥锁实验。

（12）自旋锁实验。

（13）时间转换实验。

(14) 软件定时器实验。
(15) 注册 shell 命令实验。
(16) 死锁发现实验。
(17) 调度统计实验。
(18) CPU 利用率实验。

4. 实验过程

本章所有实验的步骤大致相同，如下。
(1) 环境准备。
(2) LiteOS Studio 新建工程。
(3) 配置工程。
(4) 编辑 main.c 以及所需要的源代码文件。
(5) 编译工程。
(6) 烧写到模拟器。
(7) 运行系统并通过串口终端查看结果。

2.2 LiteOS 代码目录

1. 主函数

LiteOS 入口在工程对应的 target 目录下 main.c 中，通常 main.c 的结构和基本流程如下。

```
INT32 main(VOID)
{
    board_config();
    HardwareInit();

    PRINT_RELEASE("\n******** Hello Huawei LiteOS ******** \n"
    "\nLiteOS Kernel Version : %s\n"
    "build data : %s %s\n\n"
    "********************************* \n",
    HW_LITEOS_KERNEL_VERSION_STRING, __DATE__, __TIME__);

    UINT32 ret = OsMain();
    if (ret != LOS_OK) {
        return LOS_NOK;
    }

    OsStart();

    return 0;
}
```

该函数中，首先进行内存结束地址配置 board_config()，硬件初始化 HardwareInit()，然后打印 LiteOS 的版本信息。

接着执行 OsMain() 初始化 LiteOS 内核及例程，在 OsMain() 函数中会创建用户任务，其任务处理函数为 app_init()。

最后调用 OsStart()开始任务调度，LiteOS 开始正常工作。

在使用模拟器模拟 realview-pbx-a9 时，realview-pbx-a9 工程中的 main.c 文件代码如下所示。

```c
LITE_OS_SEC_TEXT_INIT int main(void)
{
    UINT32 ret = LOS_OK;

#ifdef __GNUC__
    ArchStackGuardInit();
#endif
    OsSetMainTask();
    OsCurrTaskSet(OsGetMainTask());

    /* 早期初始化 uart 输出 */
    uart_early_init();

    /* 系统和芯片信息 */
    osSystemInfo();

    PRINTK("\nmain core booting up...\n");
    ret = OsMain();
    if (ret != LOS_OK) {
        return LOS_NOK;
    }

#ifdef LOSCFG_KERNEL_SMP
    PRINTK("releasing %u secondary cores\n", LOSCFG_KERNEL_SMP_CORE_NUM - 1);
    release_secondary_cores();
#endif

    OsStart();

    while (1) {
        __asm volatile("wfi");
    }
}
```

2. 代码树

代码树中各级目录存放的源代码的相关说明如表 2.1 所示。

表 2.1 代码树目录结构

一级目录	二级目录	三级目录	说　　明
arch	arm	cortex_a_r	A 核架构支持
		cortex_m	M 核架构支持
	arm64		ARM64 架构支持
	csky	cskyv2	cskyv2 架构支持
	riscv	rvm32	riscv 架构支持
build			LiteOS 编译系统需要的配置及脚本
compat	cmsis		LiteOS 提供的 CMSIS-RTOS 1.0 和 2.0 接口

续表

一级目录	二级目录	三级目录	说　　明
components	ai		ai(基于 mindspore)算子库实现
	connectivity	agent_tiny	agent_tiny 端云互通组件，包括公共头文件、示例代码、客户端实现代码、操作系统适配层代码
		lwm2m	lwm2m 协议实现
		mqtt	MQTT 开源协议实现
		nb_iot	LiteOS NB-IoT API
	fs	devfs	devfs 文件系统
		fatfs	fatfs 文件系统
		kifs	kifs 文件系统
		littlefs	littlefs 文件系统
		ramfs	ramfs 文件系统
		spiffs	spiffs 文件系统
		vfs	虚拟文件系统
	gui		开源 LittlevGL 图形库
	language		语言相关组件，含 lua
	lib	cjson	C 语言 JSON 库
	log		日志等级控制
	media		媒体相关组件，含 libpng、openexif、opus、upup
	net	at_device	AT 设备适配层
		at_frame	LiteOS AT 框架 API
		ifconfig	ifconfig shell 命令实现
		los_iperf	网络带宽测试工具
		lwip/lwip_port	lwip 驱动程序及 OS 适配代码
		lwip/lwip-2.1.2	lwip 协议实现
		lwip/ppp_port	lwip 协议 ppp 端口支持
		pcap	网络抓包工具
		ping	ping shell 命令实现
		sal	socket 通信支持
		tftp_server	tftp 服务
	ota		固件升级代码
	security	mbedtls/mbedtls_port	mbed TLS 的 OS 适配代码
		mbedtls/mbedtl-2.16.8	mbed TLS 协议实现
		openssl	openssl 协议
	sensorhub	include	sensor manager 头文件
		src	sensor manager 的源码实现
	utility		解析工具，含 bidireference、curl、fastlz、freetype、harfbuzz、iconv、iniparser、json-c、jsoncpp、libxml2、sqlite、thttpd、tinyxml2

续表

一级目录	二级目录	三级目录	说　　明
demos	agenttiny_lwm2m		lwm2m 协议 demo
	agenttiny_mqtt		mqtt 协议 demo
	ai		ai 的 demo
	dtls_server		dtls 协议 demo
	fs		文件系统 demo
	gui		gui 的 demo
	ipv6_client		IPv6 协议 demo
	kernel	api	供开发者测试 LiteOS 内核的 demo 示例代码
		include	API 功能头文件存放目录
	language		语言相关组件的 demo
	lms		LMS 的 demo
	media		媒体相关组件的 demo
	nbiot_without_atiny		NB-IoT demo
	sensorhub	gyro	基于 sensorhub 传感框架定时读取 MPU6050 陀螺仪原始数据的 demo
	trace		Trace 的 demo
	utility		解析工具的 demo
doc			此目录存放的是 LiteOS 的使用文档和 API 说明等文档
include			components 各个模块所依赖的头文件
kernel	base		LiteOS 基础内核代码,包括任务、中断、软件定时器、队列、事件、信号量、互斥锁、tick 等功能
		debug	LiteOS 内核调测代码,包括队列、信号量、互斥锁及任务调度的调测
		include	LiteOS 基础内核内部使用的头文件
		mem	LiteOS 中的内存管理相关代码
		sched	任务调度支持,包括对多核的调度支持
		shellcmd	LiteOS 中与基础内核相关的 shell 命令,包括 memcheck、task、systeminfo、swtmr 等
	extended	cppsupport	C++兼容适配层底层接口
		cpup	CPU 占用率统计接口
		include	extended 目录所需的头文件
		lms	LMS(实时检测内存操作合法性算法)的库文件
		lowpower	低功耗框架相关代码
		trace	trace 事件跟踪,用于实时记录系统运行轨迹
	include		LiteOS 开源内核头文件
	init		LiteOS 内核初始化相关代码

续表

一级目录	二级目录	三级目录	说明
lib	huawei_libc		LiteOS 自研 libc 库和适配的 posix 接口
	libc		LiteOS 适配的 musl libc 库
	libsec		华为安全函数库
	zlib		开源 zlib 库
osdepends	liteos		LiteOS 提供的部分 OS 适配接口
shell		src	实现 shell 命令的代码，支持基本调试功能
		include	shell 头文件
targets	bsp		通用板级支持包
	CB2201		CB2201(ck802)开发板的开发工程源码包
	Cloud_STM32F429IGTx_FIRE		野火 STM32F429（ARM Cortex M4）开发板的开发工程源码包
	GD32E103C_START		GD32E103C_START（ARM Cortex M4）开发板的开发工程源码包
	GD32F303RGT6_BearPi		GD32F303RGT6_BearPi（ARM Cortex M4）开发板的开发工程源码包
	GD32VF103V_EVAL		GD32VF103V_EVAL（riscv）开发板的开发工程源码包
	HiFive1_Rev1_B01		HiFive1_Rev1_B01(riscv)开发板的开发工程源码包
	qemu-virt-a53		Coretex A53 的 qemu 开发工程源码包
	realview-pbx-a9		Coretex A9 的 qemu 开发工程源码包
	STM32F072_Nucleo		STM32F072_Nucleo（ARM Cortex M0）开发板的开发工程源码包
	STM32F103_FIRE_Arbitrary		野火 STM32F103（ARM Cortex M3）霸道开发板的开发工程源码包
	STM32F407_ATK_Explorer		正点原子 STM32F407（ARM Cortex M4）探索者开发板的开发工程源码包
	STM32F769IDISCOVERY		STM32F769IDISCOVERY（ARM Cortex M7）开发板的开发工程源码包
	STM32L4R9I_Discovery		STM32L4R9IDISCOVERY（ARM Cortex M4）开发板的开发工程源码包
	STM32L073_Nucleo		STM32L073_Nucleo（ARM Cortex M0＋）开发板的开发工程源码包
	STM32L431_BearPi		小熊派 STM32L431（ARM Cortex M4）开发板的开发工程源码包
	STM32L476_NB476		深创客 STM32L476（ARM Cortex M4）开发板的开发工程源码包
	STM32L496_Nucleo		STM32L496 Nucleo-144（ARM Cortex M4）开发板的开发工程源码包
	STM32L552_Nucleo		STM32L552 Nucleo（ARM Cortex M33）开发板的开发工程源码包

续表

一级目录	二级目录	三级目录	说明
tools	build		LiteOS 支持的开发板编译配置文件
	menuconfig		LiteOS 编译所需的 menuconfig 脚本
Makefile			LiteOS Makefile
.config			开发板的配置文件,如果用户不重新选择开发板,默认为野火挑战者 STM32F429 开发板的配置文件

LiteOS 系统初始任务有几种,其中,Swt_Task 是软件定时器任务,用于处理软件定时器超时回调函数。IdleCore000 是系统空闲时执行的任务。system_wq 是系统默认工作队列处理任务。SerialEntryTask 从底层 buf 读取用户的输入,初步解析命令,例如 Tab 补全、方向键等。SerialShellTask 接收命令后进一步解析,并查找匹配的命令处理函数,进行调用。

2.3 创建任务实验

1. 实验内容

本实验练习基本的任务操作方法,包含两个不同优先级任务的创建、任务延时、任务锁与解锁调度、挂起和恢复等操作,阐述任务优先级调度的机制以及各接口的应用。

2. 实验代码

打开 LiteOS 工程的 main.c 文件,在其中添加以下代码。并将以下核心代码中的 TaskDemo() 函数添加到 main() 函数中的 OsStart() 语句之后。

```
\\ ...
OsStart();

TaskDemo();
\\PRINTK("\n TaskDemo is running!\n");

\\ ...
```

核心代码如下。

```
#include "los_api_task.h"
#include "los_task.h"
#include "los_inspect_entry.h"

#define HI_TASK_PRIOR       4
#define LO_TASK_PRIOR       5
#define DELAY_INTERVAL1     5
#define DELAY_INTERVAL2     10
#define DELAY_INTERVAL3     40

STATIC UINT32 g_demoTaskHiId;
```

```c
STATIC UINT32 g_demoTaskLoId;

STATIC UINT32 HiTaskEntry(VOID)
{
    UINT32 ret;

    printf("Enter high priority task handler.\n");

    /* 延时 5 个 Tick,延时后该任务会挂起,执行剩余任务中最高优先级的任务(g_taskLoId 任务) */
    ret = LOS_TaskDelay(DELAY_INTERVAL1);
    if (ret != LOS_OK) {
        printf("Delay task failed.\n");
        return LOS_NOK;
    }

    /* 5 个 Tick 时间到了后,该任务恢复,继续执行 */
    printf("High priority task LOS_TaskDelay successfully.\n");

    /* 挂起自身任务 */
    ret = LOS_TaskSuspend(g_demoTaskHiId);
    if (ret != LOS_OK) {
        printf("Suspend high priority task failed.\n");
        ret = InspectStatusSetById(LOS_INSPECT_TASK, LOS_INSPECT_STU_ERROR);
        if (ret != LOS_OK) {
            printf("Set inspect status failed.\n");
        }
        return LOS_NOK;
    }

    printf("High priority task LOS_TaskResume successfully.\n");

    ret = InspectStatusSetById(LOS_INSPECT_TASK, LOS_INSPECT_STU_SUCCESS);
    if (ret != LOS_OK) {
        printf("Set inspect status failed.\n");
    }

    /* 删除自身任务 */
    if (LOS_TaskDelete(g_demoTaskHiId) != LOS_OK) {
        printf("Delete high priority task failed.\n");
        return LOS_NOK;
    }

    return LOS_OK;
}

/* 低优先级任务入口函数 */
STATIC UINT32 LoTaskEntry(VOID)
{
    UINT32 ret;

    printf("Enter low priority task handler.\n");
```

```c
    /* 延时10个Tick,延时后该任务会挂起,执行剩余任务中最高优先级的任务(背景任务) */
    ret = LOS_TaskDelay(DELAY_INTERVAL2);
    if (ret != LOS_OK) {
        printf("Delay low priority task failed.\n");
        return LOS_NOK;
    }

    printf("High priority task LOS_TaskSuspend successfully.\n");

    /* 恢复被挂起的任务 */
    ret = LOS_TaskResume(g_demoTaskHiId);
    if (ret != LOS_OK) {
        printf("Resume high priority task failed.\n");
        ret = InspectStatusSetById(LOS_INSPECT_TASK, LOS_INSPECT_STU_ERROR);
        if (LOS_OK != ret) {
            printf("Set inspect status failed.\n");
        }
        return LOS_NOK;
    }

    /* 删除自身任务 */
    if (LOS_TaskDelete(g_demoTaskLoId) != LOS_OK) {
        printf("Delete low priority task failed.\n");
        return LOS_NOK;
    }

    return LOS_OK;
}

/* 任务测试入口函数,创建两个不同优先级的任务 */
UINT32 TaskDemo(VOID)
{
    UINT32 ret;
    TSK_INIT_PARAM_S taskInitParam;

    printf("\nKernel task demo start to run.\n");
    /* 锁任务调度,防止新创建的任务比本任务高而发生调度 */
    LOS_TaskLock();

    /* 创建高优先级任务,由于锁任务调度,任务创建成功后不会马上执行 */
    ret = memset_s(&taskInitParam, sizeof(TSK_INIT_PARAM_S), 0, sizeof(TSK_INIT_PARAM_S));
    if (ret != EOK) {
        return ret;
    }
    taskInitParam.pfnTaskEntry = (TSK_ENTRY_FUNC)HiTaskEntry;
    taskInitParam.usTaskPrio = HI_TASK_PRIOR;
    taskInitParam.pcName = "TaskDemoHiTask";
    taskInitParam.uwStackSize = LOSCFG_BASE_CORE_TSK_DEFAULT_STACK_SIZE;
#ifdef LOSCFG_KERNEL_SMP
    taskInitParam.usCpuAffiMask = CPUID_TO_AFFI_MASK(ArchCurrCpuid());
```

```
#endif
    /* 创建高优先级任务,由于锁任务调度,任务创建成功后不会马上执行 */
    ret = LOS_TaskCreate(&g_demoTaskHiId, &taskInitParam);
    if (ret != LOS_OK) {
      LOS_TaskUnlock();
      printf("Create high priority task failed.\n");
      return LOS_NOK;
    }
    printf("Create high priority task successfully.\n");

    taskInitParam.pfnTaskEntry = (TSK_ENTRY_FUNC)LoTaskEntry;
    taskInitParam.usTaskPrio = LO_TASK_PRIOR;
    taskInitParam.pcName = "TaskDemoLoTask";
    taskInitParam.uwStackSize = LOSCFG_BASE_CORE_TSK_DEFAULT_STACK_SIZE;
    /* 创建低优先级任务,由于锁任务调度,任务创建成功后不会马上执行 */
    ret = LOS_TaskCreate(&g_demoTaskLoId, &taskInitParam);
    if (ret != LOS_OK) {
    /* 删除高优先级任务 */
      if (LOS_OK != LOS_TaskDelete(g_demoTaskHiId)) {
        printf("Delete high priority task failed.\n");
    }
    /* 解锁任务调度,此时会发生任务调度,执行就绪队列中最高优先级任务 */
    LOS_TaskUnlock();
    printf("Create low priority task failed.\n");
    return LOS_NOK;
    }
    printf("Create low priority task successfully.\n");

    /* 解锁任务调度 */
    LOS_TaskUnlock();
    LOS_TaskDelay(DELAY_INTERVAL3);

    return ret;
}
```

3. 实验结果

编译运行得到的结果如下。

```
Huawei LiteOS #
Los inspect start.

Kernel task demo start to run.
Create high priority task successfully.
Enter high priority task handler.
Create low priority task successfully.
Enter low priority task handler.
High priority task LOS_TaskDelay successfully.
High priority task LOS_TaskSuspend successfully.
High priority task LOS_TaskResume successfully.
Kernel task demo finished.
```

4. 扩展

以上创建任务的例子，可以通过把项目的.config 文件中的配置项做一修改，如下。

LOS_KERNEL_DEMO_TASK is not set

修改为

LOS_KERNEL_DEMO_TASK = y

重新编译后，可得到相同的运行结果。也可以通过图形化的 Kconfig 设置该项目。

Kconfig 是一个文本形式的文件，内核的配置菜单。.config 是编译所依据的配置。在运行 make menuconfig 后在配置界面中出现的就是 Kconfig 中的选项，在界面中看到的已经配置好的选项就是从.config 中读取出来的，当配置完成后就会将配置重新保存到.config 中，编译时 makefile 会读取.config 中配置来对内核进行编译。

在 Huawei_LiteOS 根目录下执行 make defconfig 命令会解析根目录下的.config 文件，使所有的配置项尽可能使用其默认配置，并更新.config，同时在对应的 Huawei_LiteOS 开发板工程（如 realview-pbx-a9 目录下）的子目录 include 下生成 menuconfig.h。

使用 menuconfig 前，需要 Python 环境，pip 和 kconfiglib。使用 menuconfig 配置工具前，请确保已经安装编译 LiteOS 源码的交叉编译器工具链，并加入环境变量。

为满足不同的使用场景，配置工具支持下述命令。根据使用场景，在根目录下执行下述其中一个命令即可。在执行命令前，先根据开发板复制 tools/build/config/目录下的默认配置文件 ${platform}.config 到根目录，并重命名为.config。

在 Huawei_LiteOS 根目录下执行 make menuconfig 命令会展示图形化配置界面，根据需要对系统进行配置即可。

menuconfig 的使用方式，主要包括以下方法，其中字母不区分大小写。

(1) 上下键：选择不同的行，即移动到不同的选项上。

(2) 空格键/回车：用于开启或关闭选项。

- 开启选项：对应的选项前面会显示[*]，中括号里面有一个星号，表示已经开启该选项。
- 关闭选项：对应的选项前面只显示中括号[ ]，括号里面是空。
- 如果选项后面有三个短横线加上一个右箭头，即→，表示此项下面还有子选项，按空格键/回车键后可以进入子菜单。

(3) Esc 键：返回上一级菜单，或退出 menuconfig 并提示保存。

(4) 问号？：展示配置项的帮助信息。

(5) 斜线/：进入搜索配置项界面，支持配置项的搜索。

(6) 字母 F：进入帮助模式，在界面下方会显示配置项的帮助信息，再次输入字母 F 可以退出此模式。

(7) 字母 C：进入 name 模式，在此模式下，会显示配置项对应的宏定义开关，再次输入字母 C 可以退出此模式。

(8) 字母 A：进入 all 模式，在此模式下，会展开显示菜单中的所有子选项，再次输入字母 A 可以退出此模式。

（9）字母 S：保存配置项。

（10）字母 Q：退出 menuconfig 并提示保存。

在 Huawei_LiteOS 根目录下执行 make savemenuconfig 命令会解析根目录下的.config 文件，并在对应的 Huawei_LiteOS 开发板工程的子目录 include 下生成 menuconfig.h 文件。

在 Huawei_LiteOS 根目录下执行 make defconfig 命令会解析根目录下的.config 文件，使所有的配置项尽可能使用其默认配置，并更新.config，同时在对应的 Huawei_LiteOS 开发板工程的子目录 include 下生成 menuconfig.h。

在 Huawei_LiteOS 根目录下执行 make allyesconfig 命令会解析根目录下的.config，使所有的配置项尽可能使能，即设置为 Y，并更新.config，同时在对应的 Huawei_LiteOS 开发板工程的子目录 include 下生成 menuconfig.h。

在 Huawei_LiteOS 根目录下执行 make allnoconfig 命令会解析根目录下的.config，使所有的配置项尽可能禁用，即设置为 is not set，并更新.config，同时在对应的 Huawei_LiteOS 开发板工程的子目录 include 下生成 menuconfig.h。

如果源码存在和配置项相关的错误，如未定义标识符或 LOSCFG_BASE_CORE_TSK_DEFAULT_STACK_SIZE，则需通过下列语句将该配置项包含进来，也可自行设置，如设置为 24756。

```
#include"menuconfig.h"
```

2.4 多核任务创建实验

1. 实验内容

本实验练习基本的任务操作方法，包含任务创建、任务延时、任务锁与解锁调度、挂起和恢复等操作，阐述任务优先级调度的机制以及各接口的应用。

本例创建了两个任务：TaskHi 和 TaskLo。TaskHi 为高优先级任务，绑定在当前测试任务的 CPU 上。TaskLo 为低优先级任务，不设置亲和性即不绑核。

需要在 menuconfig 菜单中完成任务模块的配置和 SMP 模式使能。

2. 实验代码

打开 LiteOS 工程的 main.c 文件，在其中添加以下代码。并将以下核心代码中的 Example_TskCaseEntry()函数添加到 main()函数中的 OsStart()语句之后。

```
\\ ...
OsStart();

Example_TskCaseEntry();
\\PRINTK("\n Task is running!\n");

\\ ...
```

核心代码如下。

```
UINT32 g_taskLoId;
```

```c
UINT32 g_taskHiId;
#define TSK_PRIOR_HI 4
#define TSK_PRIOR_LO 5

UINT32 Example_TaskHi(VOID)
{
    UINT32 ret;

    printf("[cpu %d] Enter TaskHi Handler.\r\n", ArchCurrCpuid());

    /* 延时两个Tick,延时后该任务会挂起,执行剩余任务中最高优先级的任务(g_taskLoId任务) */
    ret = LOS_TaskDelay(2);
    if (ret != LOS_OK) {
        printf("Delay Task Failed.\r\n");
        return LOS_NOK;
    }

    /* 两个Tick后,该任务恢复,继续执行 */
    printf("TaskHi LOS_TaskDelay Done.\r\n");

    /* 挂起自身任务 */
    ret = LOS_TaskSuspend(g_taskHiId);
    if (ret != LOS_OK) {
        printf("Suspend TaskHi Failed.\r\n");
        return LOS_NOK;
    }
    printf("TaskHi LOS_TaskResume Success.\r\n");
    return ret;
}

/* 低优先级任务入口函数 */
UINT32 Example_TaskLo(VOID)
{
    UINT32 ret;

    printf("[cpu %d] Enter TaskLo Handler.\r\n", ArchCurrCpuid());

    /* 延时两个Tick,延时后该任务会挂起,执行剩余任务中较高优先级的任务(背景任务) */
    ret = LOS_TaskDelay(2);
    if (ret != LOS_OK) {
        printf("Delay TaskLo Failed.\r\n");
        return LOS_NOK;
    }

    printf("TaskHi LOS_TaskDelete Success.\r\n");
    return ret;
}

/* 任务测试入口函数,创建两个不同优先级的任务 */
UINT32 Example_TskCaseEntry(VOID)
{
```

```c
    UINT32 ret;
    TSK_INIT_PARAM_S initParam = {0};

/* 锁任务调度 */
    LOS_TaskLock();

    printf("LOS_TaskLock() Success on cpu %d!\r\n", ArchCurrCpuid());

    initParam.pfnTaskEntry = (TSK_ENTRY_FUNC)Example_TaskHi;
    initParam.usTaskPrio = TSK_PRIOR_HI;
    initParam.pcName = "TaskHi";
    initParam.uwStackSize = LOSCFG_TASK_MIN_STACK_SIZE;
    initParam.uwResved = LOS_TASK_STATUS_DETACHED;
#ifdef LOSCFG_KERNEL_SMP
/* 绑定高优先级任务到 CPU1 运行 */
    initParam.usCpuAffiMask = CPUID_TO_AFFI_MASK(ArchCurrCpuid());
#endif
/* 创建高优先级任务,由于 CPU1 的调度器被锁,任务创建成功后不会马上执行 */
    ret = LOS_TaskCreate(&g_taskHiId, &initParam);
    if (ret != LOS_OK) {
        LOS_TaskUnlock();

        printf("Example_TaskHi create Failed!\r\n");
        return LOS_NOK;
    }

    printf("Example_TaskHi create Success!\r\n");

    initParam.pfnTaskEntry = (TSK_ENTRY_FUNC)Example_TaskLo;
    initParam.usTaskPrio = TSK_PRIOR_LO;
    initParam.pcName = "TaskLo";
    initParam.uwStackSize = LOSCFG_TASK_MIN_STACK_SIZE;
    initParam.uwResved    = LOS_TASK_STATUS_DETACHED;
#ifdef LOSCFG_KERNEL_SMP
/* 低优先级任务不设置 CPU 亲和性 */
    initParam.usCpuAffiMask = 0;
#endif
/* 创建低优先级任务,尽管锁任务调度,但是由于该任务没有绑定该处理器,
任务创建成功后可以马上在其他 CPU 执行 */
    ret = LOS_TaskCreate(&g_taskLoId, &initParam);
    if (ret != LOS_OK) {
        LOS_TaskUnlock();

        printf("Example TaskLo create Failed!\r\n");
        return LOS_NOK;
    }

    printf("Example_TaskLo create Success!\r\n");

/* 解锁任务调度,此时会发生任务调度,执行就绪列表中最高优先级任务 */
    LOS_TaskUnlock();
```

```
        return LOS_OK;
}
```

3. 实验结果

编译后,运行得到的结果如下。

```
LOS_TaskLock() success on cpu1!
Example_TaskHi create Success!
Example_TaskLo create Success!
[cpu2] Enter TaskLo Handler.
[cpu1] Enter TaskHi Handler.
TaskHi LOS_TaskDelete Success.
TaskHi LOS_TaskDelay Done.
```

由于 TaskLo 未设置亲和性,LOS_TaskLock 对其没有锁任务的效果,因此"Example_TaskLo create Success!"与"Enter TaskLo Handler."这两句打印并没有严格的先后顺序。

2.5 调试任务实验

1. 实验内容

本实验通过 LOS_TaskInfoGet()获取指定任务的信息,包括任务状态、优先级、任务栈大小、栈顶指针 SP、任务入口函数、已使用的任务栈大小等。

2. 实验代码

打开 LiteOS 工程的 main.c 文件,在其中添加以下代码。并将以下核心代码中的 TaskDebug()函数添加到 main()函数中的 OsStart()语句之后。

```
\\ ...
OsStart();

TaskDebug();
\\PRINTK("\n Task is running!\n");

\\ ...
```

以下为任务调试的核心代码。

```
#include "los_debug_task.h"
#include "los_task.h"


#define DELAY_INTERVAL    1000

STATIC UINT32 g_demoTaskId;
STATIC BOOL g_demodoneFlag = FALSE;
```

```c
STATIC VOID TaskEntry(VOID)
{
    INT32 i;
    UINT32 delayTime = LOS_KERNEL_TASK_DELAYINTERVAL;

    for (i = 0; i < LOS_KERNEL_TASK_CYCLE_TIMES; i++) {
        printf("Task running. Interval: %d\n", delayTime);
        LOS_TaskDelay(delayTime);
    }

    g_demodoneFlag = TRUE;
}

uint32_t TaskDebug(VOID)
{
    UINT32 ret;
    TSK_INIT_PARAM_S taskInitParam;
    TSK_INFO_S taskInfo;

    printf("\nKernel debug task start to run.\n");

    LOS_TaskLock();
    /* 创建任务 */
    ret = memset_s(&taskInitParam, sizeof(TSK_INIT_PARAM_S), 0, sizeof(TSK_INIT_PARAM_S));
    if (ret != EOK) {
        return ret;
    }
    taskInitParam.pfnTaskEntry = (TSK_ENTRY_FUNC)TaskEntry;
    taskInitParam.pcName = LOS_KERNEL_TASK_NAME;
    taskInitParam.usTaskPrio = LOS_KERNEL_TASK_PRIORITY;
    taskInitParam.uwStackSize = LOS_KERNEL_TASK_STACKSIZE;
    ret = LOS_TaskCreate(&g_demoTaskId, &taskInitParam);
    if (ret != LOS_OK) {
        LOS_TaskUnlock();
        printf("Create the task failed.\n");
        return ret;
    }
    printf("Create the task successfully.\n");

    LOS_TaskInfoGet(g_demoTaskId, &taskInfo);
    printf("\tTask information:\n\
Task name: %s\n\
Task id: %d\n\
Task status: %d\n\
Task priority: %d\n",
        taskInfo.acName, taskInfo.uwTaskID, taskInfo.usTaskStatus, taskInfo.usTaskPrio);

    printf("Start Scheduling.\n");
    LOS_TaskUnlock();

    while (!g_demodoneFlag) {
```

```
      LOS_TaskDelay(DELAY_INTERVAL);
   }
   if (LOS_TaskDelete(g_demoTaskId) != LOS_OK) {
      printf("Delete the task failed.\n");
      return ret;
   }
   printf("Delete the task successfully.\n");

   return ret;
}
```

2.6 动态内存分配实验

1. 实验内容

动态内存可以根据用户需要分配内存,本实验练习动态内存分配,包括在 menuconfig 菜单中完成动态内存的配置、初始化动态内存池、从动态内存池中申请一个内存块、在内存块中存放一个数据和打印出内存块中的数据和释放该内存块。

2. 实验代码

打开 LiteOS 工程的 main.c 文件,在其中添加以下代码。并将以下核心代码中的 DynMemDemo()函数添加到 main()函数中的 OsStart()语句之后。

```
\\ ...
        OsStart();

        DynMemDemo();
        \\PRINTK("\n Task is running!\n");

        \\...
```

核心代码如下。

```
#define MEM_USE_SIZE    4
#define MEM_USE_BUFF    828

STATIC UINT32 g_demoDynMem[MEM_DYN_SIZE / 4];

UINT32 DynMemDemo(VOID)
{
   UINT32 *mem = NULL;
   UINT32 ret;
   printf("Kernel dynamic memory demo start to run.\n");
   ret = LOS_MemInit(g_demoDynMem, MEM_DYN_SIZE);
   if (ret != LOS_OK) {
      printf("Mempool init failed.\n");
      return LOS_NOK;
   }
   printf("Mempool init successfully.\n");
```

```c
    /* 分配内存 */
    mem = (UINT32 *)LOS_MemAlloc(g_demoDynMem, MEM_USE_SIZE);
    if (mem == NULL) {
        printf("Mem alloc failed.\n");
        return LOS_NOK;
    }
    printf("Mem alloc successfully.\n");

    /* 赋值 */
    *mem = MEM_USE_BUFF;
    printf("*mem = %d.\n", *mem);

    /* 释放内存 */
    ret = LOS_MemFree(g_demoDynMem, mem);
    if (ret != LOS_OK) {
        printf("Mem free failed.\n");
        ret = InspectStatusSetById(LOS_INSPECT_DMEM, LOS_INSPECT_STU_ERROR);
        if (ret != LOS_OK) {
            printf("Set inspect status failed.\n");
        }
        return LOS_NOK;
    }
    printf("Mem free successfully.\n");
    ret = InspectStatusSetById(LOS_INSPECT_DMEM, LOS_INSPECT_STU_SUCCESS);
    if (ret != LOS_OK) {
        printf("Set inspect status failed.\n");
    }

    return LOS_OK;
}
```

3. 实验结果

输出结果如下。

```
Kernel dynamic memory demo start to run.
Mempool init successfully.
Mem alloc successfully.
*mem = 828
Mem free successfully.
```

2.7 静态内存分配实验

1. 实验内容

静态内存分配需要初始化一个静态内存池，从静态内存池中申请一块静态内存，在内存块存放一个数据，打印出内存块中的数据，清除内存块中的数据，释放该内存块等步骤。本实验练习静态内存分配。

2. 实验代码

打开 LiteOS 工程的 main.c 文件,在其中添加以下代码。并将以下核心代码中的 StaticMemDemo()函数添加到 main()函数中的 OsStart()语句之后。

```
\\ ...
    OsStart();

    StaticMemDemo();
    \\PRINTK("\n Task is running!\n");

    \\...
```

核心代码如下。

```c
#define MEM_USE_BUFF    828
#define BLOCKSIZE       3
#define POOLSIZE        144

STATIC UINT32 g_demoBoxMem[POOLSIZE];
UINT32 StaticMemDemo(VOID)
{
  UINT32 *mem = NULL;
  UINT32 ret;

  printf("Kernel static memory demo start to run.\n");
  ret = LOS_MemboxInit(&g_demoBoxMem[0], POOLSIZE, BLOCKSIZE);
  if (ret != LOS_OK) {
    printf("Mem box init failed.\n");
    return LOS_NOK;
  } else {
    printf("Mem box init successfully.\n");
  }

/* 分配内存 */
  mem = (UINT32 *)LOS_MemboxAlloc(g_demoBoxMem);
  if (mem == NULL) {
    printf("Mem box alloc failed.\n");
    return LOS_NOK;
  }
  printf("Mem box alloc successfully.\n");
/* 赋值 */
  *mem = MEM_USE_BUFF;
  printf(" *mem = %d.\n", *mem);
/* 清除内存的内容 */
  LOS_MemboxClr(g_demoBoxMem, mem);
  printf("Clear data ok, *mem = %d.\n", *mem);
/* 释放 membox */
  ret = LOS_MemboxFree(g_demoBoxMem, mem);
  if (ret == LOS_OK) {
```

```
        printf("Mem box free successfully.\n");
        ret = InspectStatusSetById(LOS_INSPECT_SMEM, LOS_INSPECT_STU_SUCCESS);
        if (ret != LOS_OK) {
          printf("Set inspect status failed.\n");
        }
    } else {
        printf("Mem box free failed.\n");
        ret = InspectStatusSetById(LOS_INSPECT_SMEM, LOS_INSPECT_STU_ERROR);
        if (ret != LOS_OK) {
          printf("Set inspect status failed.\n");
        }
    }

    return LOS_OK;
}
```

3. 实验结果

输出结果如下。

```
Kernel static memory demo start to run.
Mem box init successfully.
Mem box alloc successfully.
 *mem = 828
Clear data ok, *mem = 0
Mem box free successfully.
```

2.8 中断处理和错误处理实验

1. 实验内容

本实验用于验证创建中断、设置中断亲和性、使能中断、触发中断、屏蔽中断和删除中断的功能。在编写代码之前,需在 menuconfig 菜单中配置中断使用最大数、配置可设置的中断优先级个数。

2. 实验代码

打开 LiteOS 工程的 main.c 文件,在其中添加以下代码。并将以下核心代码中的 It_Hwi_001()函数或 TestCase()添加到 main()函数中的 OsStart()语句之后。

```
\\...
    OsStart();

    It_Hwi_001();
    \\PRINTK("\n Task is running!\n");

    \\...
```

核心代码如下。

```
#include "los_hwi.h"
```

```c
#include "los_typedef.h"
#include "los_task.h"

STATIC VOID HwiUsrIrq(VOID)
{
  printf("\n in the func HwiUsrIrq \n");
}

/* cpu0 触发, cpu0 响应 */
UINT32 It_Hwi_001(VOID)
{
  UINT32 ret;
  UINT32 irqNum = 26; /* ppi */
  UINT32 irqPri = 0x3;

  ret = LOS_HwiCreate(irqNum, irqPri, 0, (HWI_PROC_FUNC)HwiUsrIrq, 0);
  if (ret != LOS_OK) {
    return LOS_NOK;
  }

#ifdef LOSCFG_KERNEL_SMP
  ret = LOS_HwiSetAffinity(irqNum, CPUID_TO_AFFI_MASK(ArchCurrCpuid()));
  if (ret != LOS_OK) {
    return LOS_NOK;
  }
#endif
  ret = LOS_HwiEnable(irqNum);
    if (ret != LOS_OK) {
      return LOS_NOK;
    }

  ret = LOS_HwiTrigger(irqNum);
    if (ret != LOS_OK) {
      return LOS_NOK;
    }

  LOS_TaskDelay(1);

  ret = LOS_HwiDisable(irqNum);
    if (ret != LOS_OK) {
      return LOS_NOK;
    }

  ret = LOS_HwiDelete(irqNum, NULL);
  if (ret != LOS_OK) {
    return LOS_NOK;
  }

  return LOS_OK;
}
```

在下面的代码中，验证注册错误处理钩子函数和执行错误处理函数的用法。

```
#include "los_err.h"
#include "los_typedef.h"
#include <stdio.h>

void Test_ErrHandle(CHAR *fileName, UINT32 lineNo, UINT32 errorNo,
    UINT32 paraLen, VOID *para)
{
  printf("err handle ok\n");
}

static UINT32 TestCase(VOID)
{
  UINT32 errNo = 0;
  UINT32 ret;
  UINT32 errLine = 16;

  LOS_RegErrHandle(Test_ErrHandle);

  ret = LOS_ErrHandle("os_unspecific_file", errLine, errNo, 0, NULL);
  if (ret != LOS_OK) {
    return LOS_NOK;
  }

  return LOS_OK;
}
```

2.9 消息队列实验

1. 实验内容

本实验通过 LOS_QueueCreate()创建一个消息队列，通过 LOS_TaskCreate()创建任务 1 和任务 2 两个任务。任务 1 调用写队列接口发送消息，任务 2 通过读队列接口接收消息。通过 LOS_QueueDelete()删除队列。编译之前，在 menuconfig 菜单中完成队列模块的配置。

2. 实验代码

打开 LiteOS 工程的 main.c 文件，在其中添加以下代码。并将以下核心代码中的 Example_CreateTask()函数添加到 main()函数中的 OsStart()语句之后。

```
\\...
        OsStart();

        Example_CreateTask();
        \\PRINTK("\n Task is running!\n");

        \\...
```

核心代码如下。

```c
#include "los_task.h"
#include "los_queue.h"

static UINT32 g_queue;
#define BUFFER_LEN 50

VOID send_Entry(VOID)
{
    UINT32 i = 0;
    UINT32 ret = 0;
    CHAR abuf[] = "test is message x";
    UINT32 len = sizeof(abuf);

    while (i < 5) {
        abuf[len - 2] = '0' + i;
        i++;

        ret = LOS_QueueWriteCopy(g_queue, abuf, len, 0);
        if(ret != LOS_OK) {
            dprintf("send message failure, error: %x\n", ret);
        }

        LOS_TaskDelay(5);
    }
}

VOID recv_Entry(VOID)
{
    UINT32 ret = 0;
    CHAR readBuf[BUFFER_LEN] = {0};
    UINT32 readLen = BUFFER_LEN;

    while (1) {
        ret = LOS_QueueReadCopy(g_queue, readBuf, &readLen, 0);
        if(ret != LOS_OK) {
            dprintf("recv message failure, error: %x\n", ret);
            break;
        }

        dprintf("recv message: %s\n", readBuf);
        LOS_TaskDelay(5);
    }

    while (LOS_OK != LOS_QueueDelete(g_queue)) {
        LOS_TaskDelay(1);
    }

    dprintf("delete the queue success!\n");
}
```

```c
UINT32 Example_CreateTask(VOID)
{
    UINT32 ret = 0;
    UINT32 task1, task2;
    TSK_INIT_PARAM_S initParam;

    initParam.pfnTaskEntry = (TSK_ENTRY_FUNC)send_Entry;
    initParam.usTaskPrio = 9;
    initParam.uwStackSize = LOS_TASK_MIN_STACK_SIZE;
    initParam.pcName = "sendQueue";
    #ifdef LOSCFG_KERNEL_SMP
    initParam.usCpuAffiMask = CPUID_TO_AFFI_MASK(ArchCurrCpuid());
    #endif
    initParam.uwResved = LOS_TASK_STATUS_DETACHED;

    LOS_TaskLock();
    ret = LOS_TaskCreate(&task1, &initParam);
    if(ret != LOS_OK) {
        dprintf("create task1 failed, error: % x\n", ret);
        return ret;
    }

    initParam.pcName = "recvQueue";
    initParam.pfnTaskEntry = (TSK_ENTRY_FUNC)recv_Entry;
    ret = LOS_TaskCreate(&task2, &initParam);
    if(ret != LOS_OK) {
        dprintf("create task2 failed, error: % x\n", ret);
        return ret;
    }

    ret = LOS_QueueCreate("queue", 5, &g_queue, 0, BUFFER_LEN);
    if(ret != LOS_OK) {
        dprintf("create queue failure, error: % x\n", ret);
    }

    dprintf("create the queue success!\n");
    LOS_TaskUnlock();
    return ret;
}
```

3. 实验结果

输出结果如下。

```
create the queue success!
recv message: test is message 0
recv message: test is message 1
recv message: test is message 2
recv message: test is message 3
recv message: test is message 4
```

```
recv message failure, error: 200061d
delete the queue success!
```

其中,错误码 0x0200061d 表示队列已空。

2.10 事件实验

1. 实验内容

在本实验中,任务 Example_TaskEntry() 创建一个任务 Example_Event(),Example_Event() 读事件阻塞,Example_TaskEntry() 向该任务写事件。可以通过示例日志中打印的先后顺序理解事件操作时伴随的任务切换。

在任务 Example_TaskEntry() 中创建任务 Example_Event(),其中,任务 Example_Event() 优先级高于 Example_TaskEntry()。在任务 Example_Event() 中读事件 0x00000001,阻塞,发生任务切换,执行任务 Example_TaskEntry()。在任务 Example_TaskEntry() 中向任务 Example_Event() 写事件 0x00000001,发生任务切换,执行任务 Example_Event()。Example_Event() 得以执行,直到任务结束。Example_TaskEntry() 得以执行,直到任务结束。

2. 实验代码

打开 LiteOS 工程的 main.c 文件,在其中添加以下代码。并将以下核心代码中的 Example_TaskEntry() 函数添加到 main() 函数中的 OsStart() 语句之后。

```
\\ ...
    OsStart();

    Example_TaskEntry();
    \\PRINTK("\n Task is running!\n");

    \\...
```

核心代码如下。

```c
#include "los_event.h"
#include "los_task.h"
#include "securec.h"

/* 任务 ID */
UINT32 g_testTaskId;

/* 事件控制结构体 */
EVENT_CB_S g_exampleEvent;

/* 等待的事件类型 */
#define EVENT_WAIT 0x00000001

/* 用例任务入口函数 */
VOID Example_Event(VOID)
```

```c
{
    UINT32 ret;
    UINT32 event;

    /* 超时等待方式读事件,超时时间为 100 Tick, 若 100 Tick 后未读取到
    指定事件,读事件超时,任务直接唤醒 */
    printf("Example_Event wait event 0x%x \n", EVENT_WAIT);

    event = LOS_EventRead(&g_exampleEvent, EVENT_WAIT, LOS_WAITMODE_AND, 100);
    if (event == EVENT_WAIT) {
        printf("Example_Event,read event : 0x%x\n", event);
    } else {
        printf("Example_Event,read event timeout\n");
    }
}

UINT32 Example_TaskEntry(VOID)
{
    UINT32 ret;
    TSK_INIT_PARAM_S task1;

    /* 事件初始化 */
    ret = LOS_EventInit(&g_exampleEvent);
    if (ret != LOS_OK) {
        printf("init event failed .\n");
        return -1;
    }

    /* 创建任务 */
    (VOID)memset_s(&task1, sizeof(TSK_INIT_PARAM_S), 0,
      sizeof(TSK_INIT_PARAM_S));
    task1.pfnTaskEntry = (TSK_ENTRY_FUNC)Example_Event;
    task1.pcName       = "EventTsk1";
    task1.uwStackSize  = OS_TSK_DEFAULT_STACK_SIZE;
    task1.usTaskPrio   = 5;
    ret = LOS_TaskCreate(&g_testTaskId, &task1);
    if (ret != LOS_OK) {
        printf("task create failed .\n");
        return LOS_NOK;
    }

    /* 写 g_testTaskId 等待事件 */
    printf("Example_TaskEntry write event .\n");

    ret = LOS_EventWrite(&g_exampleEvent, EVENT_WAIT);
    if (ret != LOS_OK) {
        printf("event write failed .\n");
        return LOS_NOK;
    }

    /* 清标志位 */
```

```
            printf("EventMask: % d\n", g_exampleEvent.uwEventID);
            LOS_EventClear(&g_exampleEvent, ~g_exampleEvent.uwEventID);
            printf("EventMask: % d\n", g_exampleEvent.uwEventID);

            /* 删除任务 */
            ret = LOS_TaskDelete(g_testTaskId);
            if (ret != LOS_OK) {
                    printf("task delete failed .\n");
                    return LOS_NOK;
            }

            return LOS_OK;
    }
```

3. 实验结果

本程序运行后,显示如下信息。

```
Example_Event wait event 0x1
Example_TaskEntry write event.
Example_Event,read event : 0x1
EventMask: 1
EventMask: 0
```

2.11 信号量实验

1. 实验内容

在信号量实验中,首先在 menuconfig 菜单中完成信号量的配置。然后创建一个信号量,两个任务中申请同一信号量,在使用完后删除信号量。具体如下。

测试任务 Example_TaskEntry()创建一个信号量,锁任务调度,创建两个任务 Example_SemTask1()、Example_SemTask2(),Example_SemTask2()优先级高于 Example_SemTask1(),两个任务中申请同一信号量,解锁任务调度后两任务阻塞,测试任务 Example_TaskEntry()释放信号量。

Example_SemTask2()得到信号量,被调度,然后任务休眠 20Tick,Example_SemTask2()延迟,Example_SemTask1()被唤醒。Example_SemTask1()定时阻塞模式申请信号量,等待时间为 10Tick,因信号量仍被 Example_SemTask2()持有,Example_SemTask1()挂起,10Tick 后仍未得到信号量,Example_SemTask1()被唤醒,试图以永久阻塞模式申请信号量,Example_SemTask1()挂起。20Tick 后 Example_SemTask2()唤醒,释放信号量后,Example_SemTask1()得到信号量被调度运行,最后释放信号量。

Example_SemTask1()执行完,40Tick 后任务 Example_TaskEntry()被唤醒,执行删除信号量,删除两个任务。

2. 实验代码

打开 LiteOS 工程的 main.c 文件,在其中添加以下代码。并将以下核心代码中的 ExampleTaskEntry()函数添加到 main()函数中的 OsStart()语句之后。

```
\\ ...
      OsStart();

      ExampleTaskEntry();
      \\PRINTK("\n Task is running!\n");

      \\...
```

核心代码如下。

```c
#include "los_sem.h"
#include "securec.h"

/* 任务 ID */
static UINT32 g_testTaskId01;
static UINT32 g_testTaskId02;
/* 测试任务优先级 */
#define TASK_PRIO_TEST   5
/* 信号量结构体 id */
static UINT32 g_semId;

VOID Example_SemTask1(VOID)
{
    UINT32 ret;

    printf("Example_SemTask1 try get sem g_semId ,timeout 10 ticks.\n");
    /* 定时阻塞模式申请信号量,定时时间为10Tick */
    ret = LOS_SemPend(g_semId, 10);

    /* 申请到信号量 */
    if (ret == LOS_OK) {
        LOS_SemPost(g_semId);
        return;
    }
    /* 定时时间到,未申请到信号量 */
    if (ret == LOS_ERRNO_SEM_TIMEOUT) {
        printf("Example_SemTask1 timeout and try get sem
            g_semId wait forever.\n");
        /*永久阻塞模式申请信号量*/
        ret = LOS_SemPend(g_semId, LOS_WAIT_FOREVER);
        printf("Example_SemTask1 wait_forever and get
            sem g_semId.\n");
        if (ret == LOS_OK) {
            LOS_SemPost(g_semId);
            return;
        }
    }
}

VOID Example_SemTask2(VOID)
{
```

```c
        UINT32 ret;
        printf("Example_SemTask2 try get sem g_semId wait forever.\n");
        /* 永久阻塞模式申请信号量 */
        ret = LOS_SemPend(g_semId, LOS_WAIT_FOREVER);

        if (ret == LOS_OK) {
                printf("Example_SemTask2 get sem g_semId and
                    then delay20tick.\n");
        }

        /* 任务休眠 20 Tick */
        LOS_TaskDelay(20);

        printf("Example_SemTask2 post sem g_semId .\n");
        /* 释放信号量 */
        LOS_SemPost(g_semId);
        return;
}

UINT32 ExampleTaskEntry(VOID)
{
        UINT32 ret;
        TSK_INIT_PARAM_S task1;
        TSK_INIT_PARAM_S task2;

        /* 创建信号量 */
        LOS_SemCreate(0,&g_semId);

        /* 锁任务调度 */
        LOS_TaskLock();

        /* 创建任务 1 */
        (VOID)memset_s(&task1, sizeof(TSK_INIT_PARAM_S), 0,
           sizeof(TSK_INIT_PARAM_S));
        task1.pfnTaskEntry = (TSK_ENTRY_FUNC)Example_SemTask1;
        task1.pcName       = "TestTsk1";
        task1.uwStackSize  = OS_TSK_DEFAULT_STACK_SIZE;
        task1.usTaskPrio   = TASK_PRIO_TEST;
        ret = LOS_TaskCreate(&g_testTaskId01, &task1);
        if (ret != LOS_OK) {
                printf("task1 create failed .\n");
                return LOS_NOK;
        }

        /* 创建任务 2 */
        (VOID)memset_s(&task2, sizeof(TSK_INIT_PARAM_S), 0,
           sizeof(TSK_INIT_PARAM_S));
        task2.pfnTaskEntry = (TSK_ENTRY_FUNC)Example_SemTask2;
        task2.pcName       = "TestTsk2";
        task2.uwStackSize  = OS_TSK_DEFAULT_STACK_SIZE;
        task2.usTaskPrio   = (TASK_PRIO_TEST - 1);
```

```
        ret = LOS_TaskCreate(&g_testTaskId02, &task2);
        if (ret != LOS_OK) {
                printf("task2 create failed .\n");
                return LOS_NOK;
        }

        /* 解锁任务调度 */
        LOS_TaskUnlock();

        ret = LOS_SemPost(g_semId);

        /* 任务休眠 40 Tick */
        LOS_TaskDelay(40);

        /* 删除信号量 */
        LOS_SemDelete(g_semId);

        /* 删除任务 1 */
        ret = LOS_TaskDelete(g_testTaskId01);
        if (ret != LOS_OK) {
                printf("task1 delete failed .\n");
                return LOS_NOK;
        }
        /* 删除任务 2 */
        ret = LOS_TaskDelete(g_testTaskId02);
        if (ret != LOS_OK) {
                printf("task2 delete failed .\n");
                return LOS_NOK;
        }

        return LOS_OK;
}
```

3. 实验结果

编译后，运行得到的结果如下。

```
Example_SemTask2 try get sem g_semId wait forever.
Example_SemTask1 try get sem g_semId ,timeout 10 ticks.
Example_SemTask2 get sem g_semId and then delay 20ticks.
Example_SemTask1 timeout and try get sem g_semId wait forever.
Example_SemTask2 post sem g_semId.
Example_SemTask1 wait_forever and get sem g_semId.
```

2.12 互斥锁实验

1. 实验内容

在本实验中，任务 Example_TaskEntry()创建一个互斥锁，锁任务调度，创建两个任务 Example_MutexTask1()、Example_MutexTask2()。Example_MutexTask2()优先级高于

Example_MutexTask1(),解锁任务调度,然后 Example_TaskEntry()任务休眠 300Tick。

Example_MutexTask2()被调度,以永久阻塞模式申请互斥锁,并成功获取到该互斥锁,然后任务休眠 100Tick,Example_MutexTask2()挂起,Example_MutexTask1()被唤醒。

Example_MutexTask1()以定时阻塞模式申请互斥锁,等待时间为 10Tick,因互斥锁仍被 Example_MutexTask2()持有,Example_MutexTask1()挂起。10Tick 超时时间到达后,Example_MutexTask1()被唤醒,以永久阻塞模式申请互斥锁,因互斥锁仍被 Example_MutexTask2()持有,Example_MutexTask1()挂起。

100Tick 休眠时间到达后,Example_MutexTask2()被唤醒,释放互斥锁,唤醒 Example_MutexTask1()。Example_MutexTask1()成功获取到互斥锁后,释放锁。

300Tick 休眠时间到达后,任务 Example_TaskEntry()被调度运行,删除互斥锁,删除两个任务。

2. 实验代码

打开 LiteOS 工程的 main.c 文件,在其中添加以下代码。并将以下核心代码中的 Example_TaskEntry()函数添加到 main()函数中的 OsStart()语句之后。

```
\\ ...
    OsStart();

    Example_TaskEntry();
    \\PRINTK("\n Task is running!\n");

    \\...
```

核心代码如下。

```c
/* 互斥锁句柄 ID */
UINT32 g_testMux;
/* 任务 ID */
UINT32 g_testTaskId01;
UINT32 g_testTaskId02;

VOID Example_MutexTask1(VOID)
{
    UINT32 ret;

    printf("task1 try to get  mutex, wait 10 ticks.\n");
    /* 申请互斥锁 */
    ret = LOS_MuxPend(g_testMux, 10);

    if (ret == LOS_OK) {
        printf("task1 get mutex g_testMux.\n");
        /* 释放互斥锁 */
        LOS_MuxPost(g_testMux);
        return;
    } else if (ret == LOS_ERRNO_MUX_TIMEOUT ) {
        printf("task1 timeout and try to get mutex, wait forever.\n");
```

```c
                /* 申请互斥锁 */
                ret = LOS_MuxPend(g_testMux, LOS_WAIT_FOREVER);
                if (ret == LOS_OK) {
                        printf("task1 wait forever, get mutex g_testMux.\n");
                        /* 释放互斥锁 */
                        LOS_MuxPost(g_testMux);
                        return;
                }
        }
        return;
}

VOID Example_MutexTask2(VOID)
{
        printf("task2 try to get  mutex, wait forever.\n");
        /* 申请互斥锁 */
        (VOID)LOS_MuxPend(g_testMux, LOS_WAIT_FOREVER);

        printf("task2 get mutex g_testMux and suspend 100 ticks.\n");

        /* 任务休眠 100Tick */
        LOS_TaskDelay(100);

        printf("task2 resumed and post the g_testMux\n");
        /* 释放互斥锁 */
        LOS_MuxPost(g_testMux);
        return;
}

UINT32 Example_TaskEntry(VOID)
{
        UINT32 ret;
        TSK_INIT_PARAM_S task1;
        TSK_INIT_PARAM_S task2;

        /* 创建互斥锁 */
        LOS_MuxCreate(&g_testMux);

        /* 锁任务调度 */
        LOS_TaskLock();

        /* 创建任务1 */
        memset(&task1, 0, sizeof(TSK_INIT_PARAM_S));
        task1.pfnTaskEntry = (TSK_ENTRY_FUNC)Example_MutexTask1;
        task1.pcName       = "MutexTsk1";
        task1.uwStackSize  = LOSCFG_BASE_CORE_TSK_DEFAULT_STACK_SIZE;
        task1.usTaskPrio   = 5;
        ret = LOS_TaskCreate(&g_testTaskId01, &task1);
        if (ret != LOS_OK) {
                printf("task1 create failed.\n");
                return LOS_NOK;
```

```c
    }

    /* 创建任务2 */
    memset(&task2, 0, sizeof(TSK_INIT_PARAM_S));
    task2.pfnTaskEntry = (TSK_ENTRY_FUNC)Example_MutexTask2;
    task2.pcName       = "MutexTsk2";
    task2.uwStackSize  = LOSCFG_BASE_CORE_TSK_DEFAULT_STACK_SIZE;
    task2.usTaskPrio   = 4;
    ret = LOS_TaskCreate(&g_testTaskId02, &task2);
    if (ret != LOS_OK) {
            printf("task2 create failed.\n");
            return LOS_NOK;
    }

    /* 解锁任务调度 */
    LOS_TaskUnlock();
    /* 休眠300Tick */
    LOS_TaskDelay(300);

    /* 删除互斥锁 */
    LOS_MuxDelete(g_testMux);

    /* 删除任务1 */
    ret = LOS_TaskDelete(g_testTaskId01);
    if (ret != LOS_OK) {
            printf("task1 delete failed .\n");
            return LOS_NOK;
    }
    /* 删除任务2 */
    ret = LOS_TaskDelete(g_testTaskId02);
    if (ret != LOS_OK) {
            printf("task2 delete failed .\n");
            return LOS_NOK;
    }

    return LOS_OK;
}
```

3. 实验结果

编译后,运行得到的结果如下。

```
task2 try to get   mutex, wait forever.
task2 get mutex g_testMux and suspend 100 tick.
task1 try to get   mutex, wait 10 tick.
task1 timeout and try to get mutex, wait forever.
task2 resumed and post the g_testMux
task1 wait forever,get mutex g_testMux.
```

2.13 自旋锁实验

1. 实验内容

本实验中,在 menuconfig 中将 LOSCFG_KERNEL_SMP 配置项打开,并设置多核 core 数量。任务 Example_TaskEntry()初始化自旋锁,创建两个任务 Example_SpinTask1()、Example_SpinTask2(),分别运行于两个核。

Example_SpinTask1()、Example_SpinTask2()中均执行申请自旋锁的操作,同时为了模拟实际操作,在持有自旋锁后进行延迟操作,最后释放自旋锁。

300Tick 后任务 Example_TaskEntry()被调度运行,删除任务 Example_SpinTask1()和 Example_SpinTask2()。由于多核的运行时序不是固定的,因此存在任务执行顺序不同的情况。

2. 实验代码

打开 LiteOS 工程的 main.c 文件,在其中添加以下代码。并将以下核心代码中的 Example_TaskEntry()函数添加到 main()函数中的 OsStart()语句之后。

```
\\ ...
    OsStart();

    Example_TaskEntry();
    \\PRINTK("\n Task is running!\n");

    \\...
```

核心代码如下。

```c
#include "los_spinlock.h"
#include "los_task.h"

/* 自旋锁句柄 ID */
SPIN_LOCK_S g_testSpinlock;
/* 任务 ID */
UINT32 g_testTaskId01;
UINT32 g_testTaskId02;

VOID Example_SpinTask1(VOID)
{
    UINT32 i;
    UINTPTR intSave;

    /* 申请自旋锁 */
    dprintf("task1 try to get spinlock\n");
    LOS_SpinLockSave(&g_testSpinlock, &intSave);
    dprintf("task1 got spinlock\n");
    for(i = 0; i < 5000; i++) {
        asm volatile("nop");
```

```c
        }

        /* 释放自旋锁 */
        dprintf("task1 release spinlock\n");
        LOS_SpinUnlockRestore(&g_testSpinlock, intSave);

        return;
}

VOID Example_SpinTask2(VOID)
{
        UINT32 i;
        UINTPTR intSave;

        /* 申请自旋锁 */
        dprintf("task2 try to get spinlock\n");
        LOS_SpinLockSave(&g_testSpinlock, &intSave);
        dprintf("task2 got spinlock\n");
        for(i = 0; i < 5000; i++) {
                asm volatile("nop");
        }

        /* 释放自旋锁 */
        dprintf("task2 release spinlock\n");
        LOS_SpinUnlockRestore(&g_testSpinlock, intSave);

        return;
}

UINT32 Example_TaskEntry(VOID)
{
        UINT32 ret;
        TSK_INIT_PARAM_S stTask1;
        TSK_INIT_PARAM_S stTask2;

        /* 初始化自旋锁 */
        LOS_SpinInit(&g_testSpinlock);

        /* 创建任务1 */
        memset(&stTask1, 0, sizeof(TSK_INIT_PARAM_S));
        stTask1.pfnTaskEntry   = (TSK_ENTRY_FUNC)Example_SpinTask1;
        stTask1.pcName         = "SpinTsk1";
        stTask1.uwStackSize    = LOSCFG_TASK_MIN_STACK_SIZE;
        stTask1.usTaskPrio     = 5;
#ifdef LOSCFG_KERNEL_SMP
        /* 绑定任务到CPU0运行 */
        stTask1.usCpuAffiMask = CPUID_TO_AFFI_MASK(0);
#endif
        ret = LOS_TaskCreate(&g_testTaskId01, &stTask1);
        if(ret != LOS_OK) {
                dprintf("task1 create failed .\n");
```

```
                return LOS_NOK;
        }

        /* 创建任务2 */
        memset(&stTask2, 0, sizeof(TSK_INIT_PARAM_S));
        stTask2.pfnTaskEntry = (TSK_ENTRY_FUNC)Example_SpinTask2;
        stTask2.pcName       = "SpinTsk2";
        stTask2.uwStackSize  = LOSCFG_TASK_MIN_STACK_SIZE;
        stTask2.usTaskPrio   = 5;
        #ifdef LOSCFG_KERNEL_SMP
        /* 绑定任务到CPU1运行 */
        stTask1.usCpuAffiMask = CPUID_TO_AFFI_MASK(1);
        #endif
        ret = LOS_TaskCreate(&g_testTaskId02, &stTask2);
        if(ret != LOS_OK) {
                dprintf("task2 create failed .\n");
                return LOS_NOK;
        }

        /* 任务休眠300Tick */
        LOS_TaskDelay(300);

        /* 删除任务1 */
        ret = LOS_TaskDelete(g_testTaskId01);
        if(ret != LOS_OK) {
                dprintf("task1 delete failed .\n");
                return LOS_NOK;
        }
        /* 删除任务2 */
        ret = LOS_TaskDelete(g_testTaskId02);
        if(ret != LOS_OK) {
                dprintf("task2 delete failed .\n");
                return LOS_NOK;
        }

        return LOS_OK;
}
```

3. 实验结果

编译后,运行得到的结果如下。

```
task2 try to get spinlock
task2 got spinlock
task1 try to get spinlock
task2 release spinlock
task1 got spinlock
task1 release spinlock
```

2.14 时间转换实验

1. 实验内容

本实验中,实现了时间转换和时间统计等功能。时间转换可以把将毫秒数转换为 Tick 数,或将 Tick 数转换为毫秒数。时间统计可以获得每 Tick 的 Cycle 数、自系统启动以来的 Tick 数和延迟后的 Tick 数。

本实验使用每秒的 Tick 数 LOSCFG_BASE_CORE_TICK_PER_SECOND 的默认值 100。并需要根据具体的硬件设备配好 OS_SYS_CLOCK 系统主时钟频率。

2. 实验代码

打开 LiteOS 工程的 main.c 文件,在其中添加以下代码。并将以下核心代码中的 Example_TransformTime()或 Example_GetTime()函数添加到 main()函数中的 OsStart()语句之后。

```
\\ ...
        OsStart();

        Example_TransformTime();
        Example_GetTime();
        \\PRINTK("\n Task is running!\n");

        \\...
```

时间转换核心代码如下。

```
VOID Example_TransformTime(VOID)
{
        UINT32 ms;
        UINT32 tick;

        tick = LOS_MS2Tick(10000);            //将 10000ms 转换为 Tick
        dprintf("tick = %d \n",tick);
        ms = LOS_Tick2MS(100);                //100Tick 转换为 ms
        dprintf("ms = %d \n",ms);
}
```

时间统计和时间延迟核心代码如下。

```
VOID Example_GetTime(VOID)
{
        UINT32 cyclePerTick;
        UINT64 tickCount;

        cyclePerTick  = LOS_CyclePerTickGet();
        if(0 != cyclePerTick) {
                dprintf("LOS_CyclePerTickGet = %d \n", cyclePerTick);
        }
```

```
        tickCount = LOS_TickCountGet();
        if(0 != tickCount) {
                dprintf("LOS_TickCountGet = %d \n", (UINT32)tickCount);
        }
        LOS_TaskDelay(200);
        tickCount = LOS_TickCountGet();
        if(0 != tickCount) {
                dprintf("LOS_TickCountGet after delay = %d \n", (UINT32)tickCount);
        }
}
```

3. 实验结果

第 1 段程序编译运行得到的结果为给出 Tick 数目和毫秒数目。

```
tick = 1000
ms = 1000
```

第 2 段程序编译运行得到的结果如下。

```
LOS_CyclePerTickGet = 495000
LOS_TickCountGet = 1
LOS_TickCountGet after delay = 201
```

2.15 软件定时器实验

1. 实验内容

本实验练习 LiteOS 软件定时器的创建、启动、停止、删除等操作。熟悉单次软件定时器、周期软件定时器使用方法。在此之前，需要在 menuconfig 菜单中完成软件定时器的配置。

2. 实验代码

打开 LiteOS 工程的 main.c 文件，在其中添加以下代码。并将以下核心代码中的 Timer_example() 函数添加到 main() 函数中的 OsStart() 语句之后。

```
\\ ...
        OsStart();

        Timer_example();
        \\PRINTK("\n Task is running!\n");

        \\ ...
```

核心代码如下。

```
UINT32 g_timerCount1 = 0;
UINT32 g_timerCount2 = 0;
```

```c
VOID Timer1_CallBack(UINT32 arg)
{
    UINT64 lastTick;

    g_timerCount1++;
    lastTick = (UINT32)LOS_TickCountGet();
    dprintf("g_timerCount1 = %d\n", g_timerCount1);
    dprintf("tick_last1 = %d\n", lastTick);
}

VOID Timer2_CallBack(UINT32 arg)
{
    UINT64 lastTick;

    lastTick = (UINT32)LOS_TickCountGet();
    g_timerCount2++;
    dprintf("g_timerCount2 = %d\n", g_timerCount2);
    dprintf("tick_last2 = %d\n", lastTick);
}

VOID Timer_example(VOID)
{
    UINT16 id1;                          //Timer1 ID
    UINT16 id2;                          //Timer2 ID
    UINT32 tick;

    LOS_SwtmrCreate(1000, LOS_SWTMR_MODE_ONCE, Timer1_CallBack, &id1, 1);
    LOS_SwtmrCreate(100, LOS_SWTMR_MODE_PERIOD, Timer2_CallBack, &id2, 1);
    dprintf("create Timer1 success\n");

    LOS_SwtmrStart(id1);
    dprintf("start Timer1 sucess\n");
    LOS_TaskDelay(200);
    LOS_SwtmrTimeGet(id1, &tick);
    dprintf("tick = %d\n", tick);
    LOS_SwtmrStop(id1);
    dprintf("stop Timer1 sucess\n");

    LOS_SwtmrStart(id1);
    LOS_TaskDelay(1000);
    LOS_SwtmrDelete(id1);
    dprintf("delete Timer1 sucess\n");

    LOS_SwtmrStart(id2);
    dprintf("start Timer2\n");
    LOS_TaskDelay(1000);
    LOS_SwtmrStop(id2);
    LOS_SwtmrDelete(id2);
}
```

3. 实验结果

编译后,运行得到的结果如下。

```
Create Timer1 success
start Timer1 sucess
tick = 800
Stop Timer1 sucess
g_timerCount1 = 1201
tick_last1 = 1201
delete Timer1 success
Start Timer2
g_timerCount2 = 1
tick_last2 = 1301
g_timerCount2 = 2
tick_last2 = 1401
g_timerCount2 = 3
tick_last2 = 1501
g_timerCount2 = 4
tick_last2 = 1601
g_timerCount2 = 5
tick_last2 = 1701
g_timerCount2 = 6
tick_last2 = 1801
g_timerCount2 = 7
tick_last2 = 1901
g_timerCount2 = 8
tick_last2 = 2001
g_timerCount2 = 9
tick_last2 = 2101
g_timerCount2 = 10
tick_last2 = 2201
```

2.16 注册 shell 命令实验

1. 实验内容

本实验进行 shell 命令的注册练习,包括静态 shell 命令和动态 shell 命令。

2. 实验代码

打开 LiteOS 工程的 main.c 文件,在其中添加以下静态注册和动态注册 AT 指令的代码。

核心代码如下。

```
#include "shell.h"                    //声明动态注册函数的头文件
#include "shcmd.h"                    //声明静态注册函数的头文件

int cmd_test(void)
{
```

```
        printf("I am static test shell!\n");
        return 0;
}

int cmd_testd(void)
{
        printf("I am dynamic test shell!\n");
        return 0;
}

//静态注册
SHELLCMD_ENTRY(test_shellcmd, CMD_TYPE_EX, "test", 0, (CMD_CBK_FUNC)cmd_test);
//动态注册
void app_init(void)
{
        ....
        osCmdReg(CMD_TYPE_EX, "testd", 0, (CMD_CBK_FUNC) cmd_testd);
        ....
}
```

在链接选项中添加链接该新增命令项参数,即在 build/mk/liteos_tables_ldflags.mk 中 LITEOS_TABLES_LDFLAGS 项下添加-utest_shellcmd。

```
LITEOS_TABLES_LDFLAGS += -utest_shellcmd
```

通过 make menuconfig 使能 shell,即设置 LOSCFG_SHELL=y。

3. 实验结果

编译烧录系统后,可以执行新增的 shell 命令。

```
Huawei LiteOS # test
I am static test shell!

Huawei LiteOS # testd
I am dynamic test shell!
```

2.17 死锁发现实验

1. 实验内容

本实验通过 shell 命令 dlock 发现互斥锁死锁。任务发生死锁后,无法得到调度,通过记录任务上次调度的时间,设置一个超时时间阈值,如果任务在这段时间内都没有得到调度,则怀疑该任务发生了死锁。

首先构造 ABBA 互斥锁死锁场景。假设有两个任务,为 app_Task 和 mutexDlock_Task,同时系统中还存在其他系统默认初始任务。在任务 app_Task 中执行 MutexDlockDebug() 函数,并在该函数中创建任务 mutexDlock_Task。函数 MutexDlockDebug()(即任务 app_Task)创建了个互斥锁 mutex0,并持有 mutex0,接着创建更高优先级的任务 mutexDlock_Task,休眠一段时间后去申请 mutex1 被阻塞(任务

mutexDlock_Task 已经率先持有 mutex1)。任务 mutexDlock_Task 创建并持有 mutex1，然后申请 mutex0 被阻塞(任务 app_Task 已经率先持有 mutex0)。

2. 实验代码

打开 LiteOS 工程的 main.c 文件，在其中添加以下代码。并将以下核心代码中的 DlockDebugTask()函数添加到 main()函数中的 OsStart()语句之后。

```
\\ ...
        OsStart();

        DlockDebugTask();
        \\PRINTK("\n Task is running!\n");

        \\...
```

核心代码如下。

```
#include "unistd.h"
#include "los_mux.h"
#include "los_task.h"

static UINT32 mutexTest[2];
extern UINT32 OsShellCmdMuxDeadlockCheck(UINT32 argc, const CHAR ** argv);

VOID DlockDebugTask(VOID)
{
        UINT32 ret;

        ret = LOS_MuxPend(mutexTest[1], LOS_WAIT_FOREVER);
        if (ret != LOS_OK) {
                PRINT_ERR("pend mutex1 error % u\n", ret);
        }

        ret = LOS_MuxPend(mutexTest[0], LOS_WAIT_FOREVER);
        if (ret != LOS_OK) {
                PRINT_ERR("pend mutex0 error % u\n", ret);
        }

        ret = LOS_MuxPost(mutexTest[1]);
        if (ret != LOS_OK) {
                PRINT_ERR("post mutex1 error % u\n", ret);
        }

        ret = LOS_MuxPost(mutexTest[0]);
        if (ret != LOS_OK) {
                PRINT_ERR("post mutex0 error % u\n", ret);
        }
}

//MutexDlockDebug()函数在用户任务 app_Task 中被调度
STATIC UINT32 MutexDlockDebug(VOID)
```

```c
{
    UINT32 ret;
    UINT32 taskId;
    TSK_INIT_PARAM_S debugTask;

    ret = LOS_MuxCreate(&mutexTest[0]);
    if (ret != LOS_OK) {
        PRINT_ERR("create mutex0 error %u\n", ret);
    }

    ret = LOS_MuxCreate(&mutexTest[1]);
    if (ret != LOS_OK) {
        PRINT_ERR("create mutex1 error %u\n", ret);
    }

    ret = LOS_MuxPend(mutexTest[0], LOS_WAIT_FOREVER);
    if (ret != LOS_OK) {
        PRINT_ERR("pend mutex0 error %u\n", ret);
    }

    (VOID)memset_s(&debugTask, sizeof(TSK_INIT_PARAM_S), 0, sizeof(TSK_INIT_PARAM_S));
    debugTask.pfnTaskEntry = (TSK_ENTRY_FUNC)DlockDebugTask;
    debugTask.uwStackSize = LOSCFG_BASE_CORE_TSK_DEFAULT_STACK_SIZE;
    debugTask.pcName = "mutexDlock_Task";
    debugTask.usTaskPrio = 9;
    debugTask.uwResved = LOS_TASK_STATUS_DETACHED;

    ret = LOS_TaskCreate(&taskId, &debugTask);   /* 创建 mutexDlock_Task 任务,任务入
                                                    口函数 DlockDebugTask(),优先级 9
                                                    高于 app_Task 任务 */
    if (ret != LOS_OK) {
        PRINT_ERR("create debugTask error %u\n", ret);
    }

    sleep(2);
    ret = LOS_MuxPend(mutexTest[1], LOS_WAIT_FOREVER);
    if (ret != LOS_OK) {
        PRINT_ERR("pend mutex1 error %u\n", ret);
    }

    ret = LOS_MuxPost(mutexTest[0]);
    if (ret != LOS_OK) {
        PRINT_ERR("post mutex0 error %u\n", ret);
    }

    ret = LOS_MuxPost(mutexTest[1]);
    if (ret != LOS_OK) {
        PRINT_ERR("post mutex1 error %u\n", ret);
    }
    return ret;
}
```

配置宏 LOSCFG_DEBUG_DEADLOCK，即在 menuconfig 配置项中开启"Enable Mutex Deadlock Debugging"，开启死锁检测功能。

3．实验结果

死锁检测输出的是超过时间阈值（默认为 10min）的任务信息，编译运行后，在 shell 中运行 dlock 命令检测死锁。

```
Huawei LiteOS # dlock
Start mutexs deadlock check:
Task_name: SendToSer, ID: 0x0, holds the Mutexs below:
null
 ****** backtrace begin *******
 ******* backtrace end ********

Task_name: WowWriteFlashTask, ID: 0x3, holds the Mutexs below:
null
 ****** backtrace begin *******
 ******* backtrace end ********

Task_name: system_wq, ID: 0x4, holds the Mutexs below:
null
 ****** backtrace begin *******
 ******* backtrace end ********

Task_name: app_Task, ID: 0x5, holds the Mutexs below:
< Mutex0 info >
Ptr handle: 0x8036104c
Owner: app_Task
Count: 1
Pended task: 0. name: mutexDlock_Task, id: 0xc
 ****** backtrace begin *******
 ******* backtrace end ********

Task_name: Swt_Task, ID: 0x6, holds the Mutexs below:
null
 ****** backtrace begin *******
 ****** backtrace begin *******
traceback 0 -- lr = 0x4    fp = 0x0
 ******* backtrace end ********

Task_name: IdleCore000, ID: 0x7, holds the Mutexs below:
null
 ****** backtrace begin *******
 ******* backtrace end ********

Task_name: eth_irq_Task, ID: 0xb, holds the Mutexs below:
null
 ****** backtrace begin *******
 ******* backtrace end ********
```

```
Task_name: mutexDlock_Task, ID: 0xc, holds the Mutexs below:
<Mutex0 info>
Ptr handle: 0x80361060
Owner: mutexDlock_Task
Count: 1
Pended task: 0. name: app_Task        , id: 0x5
******* backtrace begin *******
******* backtrace end *******

----------- End -----------
```

"Task_name：app_Task，ID：0x5，holds the Mutexs below：" 和 "Task_name：mutexDlock_Task，ID：0xc，holds the Mutexs below：" 这两行后面有 mutex 信息，表示可能是任务 app_Task（任务 ID 为 0x5）和 mutexDlock_Task（任务 ID 为 0xc）发生了死锁。

在 shell 中运行 task 命令显示当前所有正在运行的任务状态和信息。找到疑似发生死锁的任务 app_Task 和 mutexDlock_Task 的入口地址。

打开 /out/<platform> 目录下 .asm 反汇编文件，找到相应的地址，即可定位到互斥锁 pend 的位置及调用的接口。

2.18 调度统计实验

1. 实验内容

本实验练习统计 CPU 的任务调度信息，通过 shell 命令进行。过程如下：

（1）通过 menuconfig 开启该调度统计功能，即配置 LOSCFG_DEBUG_SCHED_STATISTICS=y。

（2）编写程序将调度统计功能开启函数 OsShellStatisticsStart()注册为 shell 命令 mpstart、调度统计功能关闭函数 OsShellStatisticsStop()注册为 shell 命令 mpstop、显示 CPU 调度信息的函数 OsShellCmdDumpSched()注册为 shell 命令 mpstat。源代码参考 2.16 节编写。

（3）通过 menuconfig 开启该调度统计功能，即配置 LOSCFG_DEBUG_SCHED_STATISTICS=y，该功能默认关闭，菜单路径为 Debug -> Enable a Debug Version -> Enable Debug LiteOS Kernel Resource -> Enable Scheduler Statistics Debugging。

（4）编译代码、烧写运行。

2. 实验结果

在 shell 窗口中调用注册的命令查看调度信息。

调用 mpstart 命令开启调度统计，调用 mpstop 后输出调度统计信息，如图 2.1 所示。

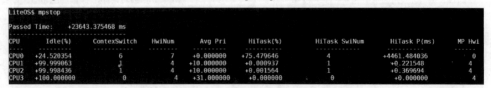

图 2.1 调度统计信息

调用 mpstat 命令后输出 CPU 调度信息，如图 2.2 所示。

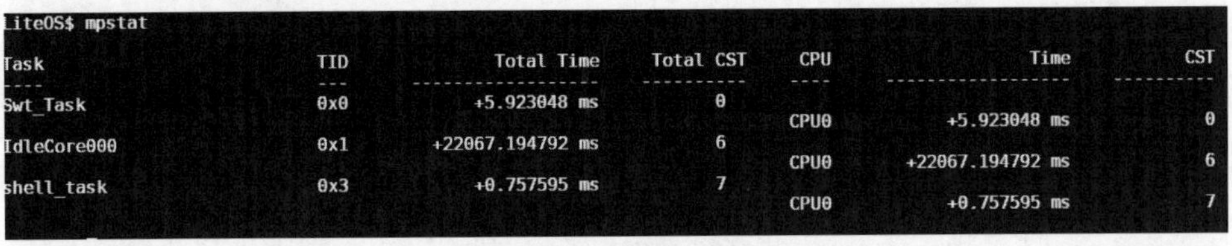

图 2.2 CPU 调度信息

2.19 CPU 利用率实验

1. 实验内容

CPU 利用率可以通过 shell 命令 CPUP 获得，也可通过代码获取。通过 CPUP 命令，能够显示系统最近 10s 的 CPU 利用率。可以指定参数显示最近 1s 的 CPU 利用率和系统启动后总的 CPU 利用率。需使能 CPU 占用率模块，通过 make menuconfig 使能，路径为 Kernel -> Enable Extend Kernel -> Enable Cpup。

CPU 利用率实验用于了解任务或系统的 CPU 利用率，实现如下功能。

（1）创建一个测试 CPU 利用率的任务。
（2）获取系统最近 1s 内所有任务或中断的 CPU 利用率。
（3）获取除 idel 任务外的系统最近 10s 内的总 CPU 利用率。
（4）获取 CPU 利用率测试任务的 CPU 利用率。

2. 实验代码

打开 LiteOS 工程的 main.c 文件，在其中添加以下代码。并将以下核心代码中的 Example_CPUP_Test()函数添加到 main()函数中的 OsStart()语句之后。

```
\\ ...
    OsStart();

    Example_CPUP_Test();
    \\PRINTK("\n Task is running!\n");

    \\...
```

核心代码如下。

```
#include <unistd.h>
#include "los_task.h"
#include "los_cpup.h"

#define MAXTASKNUM    32

UINT32 cpupUse;
UINT32 g_cpuTestTaskId;
```

```c
VOID Example_CPUP(VOID)
{
        printf("entry cpup test example\n");
        while(1) {
                usleep(100);
        }
}

UINT32 Example_CPUP_Test(VOID)
{
        UINT32 ret;
        TSK_INIT_PARAM_S cpupTestTask;
        CPUP_INFO_S cpupInfo;

        /* 创建测试 CPU 利用率的任务 */
        memset(&cpupTestTask, 0, sizeof(TSK_INIT_PARAM_S));
        cpupTestTask.pfnTaskEntry = (TSK_ENTRY_FUNC)Example_CPUP;
        cpupTestTask.pcName       = "TestCpupTsk";      /* 测试任务名称 */
        cpupTestTask.uwStackSize  = LOSCFG_BASE_CORE_TSK_DEFAULT_STACK_SIZE;
        cpupTestTask.usTaskPrio   = 5;
        cpupTestTask.uwResved     = LOS_TASK_STATUS_DETACHED;

        ret = LOS_TaskCreate(&g_cpuTestTaskId, &cpupTestTask);
        if(ret != LOS_OK) {
                printf("cpupTestTask create failed.\n");
                return LOS_NOK;
        }
        usleep(100);

        /* 系统中运行的任务或者中断数量 */
        UINT16 maxNum = MAXTASKNUM;

        /* 获取系统所有任务或中断最近 1s 的 CPU 利用率 */
        cpupUse = LOS_AllCpuUsage(maxNum, &cpupInfo, CPUP_LAST_ONE_SECONDS, 0);
        printf("the system cpu usage in last 1s: %d\n", cpupUse);

        /* 获取最近 10s 内系统(除 idel 任务外)总 CPU 利用率 */
        cpupUse = LOS_HistorySysCpuUsage(CPUP_LAST_TEN_SECONDS);
        printf("the history system cpu usage in last 10s: %d\n", cpupUse);

        /* 获取指定任务在最近 1s 的 CPU 利用率,该测试例程中指定的任务为上面创建的 CPU 利用率测试任务 */
        cpupUse = LOS_HistoryTaskCpuUsage(g_cpuTestTaskId, CPUP_LAST_ONE_SECONDS);
        printf("cpu usage of the cpupTestTask in last 1s:\n TaskID: %d\n usage: %d\n", g_cpuTestTaskId, cpupUse);

        return LOS_OK;
}
```

3. 实验结果

编译运行得到的结果如下。注意计量单位是千分位,最近 1s 系统 CPU 利用率是

1.5%。

```
the system cpu usage in last 1s: 15
the history system cpu usage in last 10s: 3
cpu usage of the cpupTestTask in last 1s:
TaskID: 10
usage: 0
```

第3章 轻量级系统设备开发实验

一个人想做点事业,非得走自己的路。

要开创新路子,最关键的是你会不会自己提出问题,能正确地提出问题就是迈开了创新的第一步。

——李政道

3.1 轻量级系统设备开发实验概览

1. 实验目的

本章实验的目的是掌握 OpenHarmony 轻量级系统设备开发的基本功能,包括 GPIO 输入输出、PWM 输出、I2C 输入输出、AT 指令、WiFi 连接和 MQTT 传输数据等。

2. 实验设备

(1) 硬件设备包含一台主流配置的计算机,以及 HiSpark 的 WiFi IoT 套件中的 Hi3861 核心板和环境监测板。

(2) 软件环境主要是 Windows 10 操作系统、DevEco Device Tool 3.0 和 OpenHarmony 3.0 LTS 等。详细的环境配置和安装过程请参考第 1 章。

3. 实验内容

包含 8 个实验,分别为:

(1) Hi3861 GPIO 输出实验。
(2) Hi3861 GPIO 查询方式输入实验。
(3) Hi3861 GPIO 中断方式输入实验。
(4) Hi3861 PWM 输出实验。
(5) Hi3861 I2C 读取 AHT 实验。
(6) Hi3861 AT 指令实验。
(7) Hi3861 WiFi 连接实验。
(8) Hi3861 MQTT 客户端实验。

4. 实验过程

本章所有实验遵循以下步骤。

(1) 环境准备。
(2) 工程配置。
(3) 在 OpenHarmony SDK 相关目录新建源文件。
(4) 构建编译环境。
(5) 编译工程。
(6) 烧写到 Hi3861 核心板。
(7) 运行系统。
(8) 观察实验板现象以及通过 DevEco Device Tool 串口终端查看结果。

OpenHarmony SDK 源码的主要目录如表 3.1 所示。

表 3.1 源码的目录

目 录 名	描 述
applications	应用程序样例，包括 camera 等
ark	方舟编译
base	基础软件服务子系统集 & 硬件服务子系统集
build	组件化编译、构建和配置脚本
docs	说明文档
domains	增强软件服务子系统集
drivers	驱动子系统
foundation	系统基础能力子系统集
kernel	内核子系统
prebuilts	编译器及工具链子系统
test	测试子系统
third_party	开源第三方组件
utils	常用的工具集
vendor	厂商提供的软件
build.py	编译脚本文件

3.2 Hi3861 GPIO 输出实验

1. 实验内容

本实验是在 Hi3861 核心板上实现通过 GPIO 控制 LED 灯的功能。

2. 实验代码

在 OpenHarmony 3.0 LTS 版本中，在\applications\sample\wifi-iot\app 目录下创建 led_demo 目录，在 led.demo 目录下创建名为 led_example.c 的文件。

通过 GPIO 控制 LED 的 led_example.c 文件完整代码如下。

```
#include <stdio.h>
#include <unistd.h>
#include "ohos_init.h"
#include "cmsis_os2.h"
```

```c
#include "iot_gpio.h"

#define LED_INTERVAL_TIME_US 300000
#define LED_TASK_STACK_SIZE 512
#define LED_TASK_PRIO 25
/* 实现IOT外设控制,首先需要通过查阅原理图明确接线关系.经过查阅,发现hispark
pegasus的LED与芯片的9号管脚相连 */
#define LED_TEST_GPIO 9 //for hispark_pegasus
enum LedState {
    LED_ON = 0,
    LED_OFF,
    LED_SPARK,
};

//在循环任务中通过周期性亮灭形式实现LED闪烁
enum LedState g_ledState = LED_SPARK;

static void * LedTask(const char * arg)
{
    (void)arg;
    while (1) {
        switch (g_ledState) {
            case LED_ON:
            IoTGpioSetOutputVal(LED_TEST_GPIO, 0);
            usleep(LED_INTERVAL_TIME_US);
            break;
            case LED_OFF:
            IoTGpioSetOutputVal(LED_TEST_GPIO, 1);
            usleep(LED_INTERVAL_TIME_US);
            break;
            case LED_SPARK:
            IoTGpioSetOutputVal(LED_TEST_GPIO, 0);
            usleep(LED_INTERVAL_TIME_US);
            IoTGpioSetOutputVal(LED_TEST_GPIO, 1);
            usleep(LED_INTERVAL_TIME_US);
            break;
            default:
            usleep(LED_INTERVAL_TIME_US);
            break;
        }
    }

    return NULL;
}

static void LedExampleEntry(void)
{
    osThreadAttr_t attr;

    /* 管脚初始化 */
    IoTGpioInit(LED_TEST_GPIO);
```

```
    /* 配置 9 号管脚为输出方向 */
    IoTGpioSetDir(LED_TEST_GPIO, IOT_GPIO_DIR_OUT);

    attr.name = "LedTask";
    attr.attr_bits = 0U;
    attr.cb_mem = NULL;
    attr.cb_size = 0U;
    attr.stack_mem = NULL;
    attr.stack_size = LED_TASK_STACK_SIZE;
    attr.priority = LED_TASK_PRIO;

    /* 启动任务 */
    if (osThreadNew((osThreadFunc_t)LedTask, NULL, &attr) == NULL) {
        printf("[LedExample] Falied to create LedTask!\n");
    }
}

SYS_RUN(LedExampleEntry);
```

在包含示例程序 led_example.c 的目录下，创建 BUILD.gn 文件，文件内容如下。

```
//创建名为 led_demo 的静态库
static_library("led_example") {
    //编译该静态库需要的源代码文件列表
    sources = [
    "led_example.c"
    ]
    //编译该静态库需要的包含目录列表
    include_dirs = [
    "//utils/native/lite/include",//ohos_init.h 所在目录
    "//kernel/liteos_m/kal/cmsis",//cmsis_os 2.h 所在目录
    "//base/iot_hardware/peripheral/interfaces/kits",//iot_gpio.h 所在目录
    ]
}
```

在上级目录下，修改 applications/sample/wifi-iot/app/BUILD.gn 文件，使 led_example.c 参与编译。

```
import("//build/lite/config/component/lite_component.gni")
    lite_component("app") {
        features = [
        "iothardware:led_example"
        ]
    }
```

3. 实验结果

烧录完成后。复位 Hi3861 核心板，板载 LED 灯不停闪烁，如图 3.1 所示。

4. 扩展

Hi3861 的 GPIO 接口具有以下功能特点。

（1）时钟源可选择：工作模式晶体时钟频率 24MHz/40MHz、低功耗模式 32kHz 时钟

图 3.1 GPIO 控制 LED 灯闪烁

频率。

(2) 1 组 GPIO,共 15 个独立的可配置管脚。

(3) 每个 GPIO 管脚都可单独控制传输方向。

(4) 每个 GPIO 可以单独被配置为外部中断源。

(5) GPIO 用作中断时有 4 种中断触发方式,中断时触发方式可配:上升沿触发,下降沿触发,高电平触发,低电平触发。

(6) GPIO 上报一个中断,CPU 查询上报的 GPIO 编号。

(7) 中断支持独立屏蔽的功能,脉冲中断支持可清除功能。

3.3 Hi3861 GPIO 查询方式输入实验

1. 实验内容

本实验是在 Hi3861 核心板上,完成通过 GPIO 实现按键对 LED 灯的控制功能。

2. 实验代码

在 OpenHarmony 3.0 LTS 版本中,在\applications\sample\wifi-iot\app 目录下创建 iotkey 目录,在 iotkey 目录下创建名为 keyPolling.c 的文件。

通过按键控制 LED 的 keyPolling.c 文件完整代码列表如下。

```
#include <stdio.h>

#include <unistd.h>
#include "ohos_init.h"
#include "cmsis_os2.h"
#include "iot_gpio.h"
#include "iot_gpio_ex.h"

#define LED_INTERVAL_TIME_US 300000
#define LED_TASK_STACK_SIZE 4096
```

```c
#define LED_TASK_PRIO 25
#define LED_TEST_GPIO 9 //for hispark_pegasus
#define KEY_GPIO 5
enum KeyState {
    KEY_DOWN = 0,
    KEY_UP,
};
enum LedState {
    LED_ON = 0,
    LED_OFF,
    LED_SPARK,
};

enum LedState g_ledState = LED_SPARK;

static void * KeyPollingTask(const char * arg)
{
    (void)arg;
    printf("\n Task is running!\n");
    while (1) {
        IotGpioValue value = IOT_GPIO_VALUE0;
        IoTGpioGetInputVal(KEY_GPIO, &value);
        if (value == KEY_UP){
            switch (g_ledState) {
                case LED_ON:
                IoTGpioSetOutputVal(LED_TEST_GPIO, 0);
                usleep(LED_INTERVAL_TIME_US);
                break;
                case LED_OFF:
                IoTGpioSetOutputVal(LED_TEST_GPIO, 1);
                usleep(LED_INTERVAL_TIME_US);
                break;
                case LED_SPARK:
                IoTGpioSetOutputVal(LED_TEST_GPIO, 0);
                usleep(LED_INTERVAL_TIME_US);
                IoTGpioSetOutputVal(LED_TEST_GPIO, 1);
                usleep(LED_INTERVAL_TIME_US);
                break;
                default:
                usleep(LED_INTERVAL_TIME_US);
                break;
            }
        }

    }

    return NULL;
}

static void KeyPollingEntry(void)
{
```

```c
        osThreadAttr_t attr;

        IoTGpioInit(LED_TEST_GPIO);
        IoTGpioSetDir(LED_TEST_GPIO, IOT_GPIO_DIR_OUT);

        IoTGpioDeinit(KEY_GPIO);
        IoTGpioInit(KEY_GPIO);
        IoSetFunc(IOT_IO_NAME_GPIO_5, IOT_IO_FUNC_GPIO_5_GPIO);
        IoTGpioSetDir(KEY_GPIO, IOT_GPIO_DIR_IN);
        IoSetPull(KEY_GPIO,1);

        attr.name = "KeyPollingTask";
        attr.attr_bits = 0U;
        attr.cb_mem = NULL;
        attr.cb_size = 0U;
        attr.stack_mem = NULL;
        attr.stack_size = LED_TASK_STACK_SIZE;
        attr.priority = LED_TASK_PRIO;

        if (osThreadNew((osThreadFunc_t)KeyPollingTask, NULL, &attr) == NULL) {
                printf("[LedExample] Falied to create LedTask!\n");
        }
}

SYS_RUN(KeyPollingEntry);
```

在包含示例程序 keyPolling.c 的目录下,创建 BUILD.gn 文件,文件内容如下。

```
//定义名为 iotkey 的静态库
static_library("iotkey") {
        //编译该静态库需要的源代码文件列表
        sources = [
        "keyPolling.c"
        ]
        //编译该静态库需要的包含目录列表
        include_dirs = [
        "//utils/native/lite/include",
        "//kernel/liteos_m/kal/cmsis",
        "//base/iot_hardware/peripheral/interfaces/kits",
        ]
}
```

在上级目录下,修改 applications/sample/wifi-iot/app/BUILD.gn 文件,使 keyPolling.c 参与编译。

```
import("//build/lite/config/component/lite_component.gni")
        lite_component("app") {
                features = [
                "iotkey: iotkey"
                ]
        }
```

3. 实验结果

烧录完成后。复位 Hi3861 核心板，在 DevEco Device Tool 的串口监视窗口，输出如下内容。

```
ready to OS start
sdk ver: Hi3861V100R001C00SPC025 2020-09-03 18:10:00
FileSystem mount ok.
wifi init success!
hilog will init.
hievent will init.
hievent init success.

Task is doing!
hiview init success.No crash dump found!
```

同时板载 LED 灯不停闪烁，按下 S2 按键时 LED 灯停止闪烁。

3.4 Hi3861 GPIO 中断方式输入实验

1. 实验内容

本实验是在 Hi3861 核心板上，完成通过 GPIO 实现按键中断事件对 LED 灯的控制功能。

2. 实验代码

在 OpenHarmony 3.0 LTS 版本中，在\applications\sample\wifi-iot\app 目录下创建 iotkey 目录，在 iotkey 目录下创建名为 keyInt.c 的文件。

通过按键中断事件控制 LED 的 keyInt.c 文件完整代码如下。

```c
#include <stdio.h>

#include <unistd.h>

#include "ohos_init.h"
#include "cmsis_os2.h"
#include "wifiiot_gpio.h"
#include "wifiiot_gpio_ex.h"

#define LED_INTERVAL_TIME_US 300000
#define LED_TASK_STACK_SIZE 512
#define LED_TASK_PRIO 25
#define LED_TEST_GPIO 9 //for hispark_pegasus
#define KEY_GPIO 5
enum KeyState {
    KEY_DOWN = 0,
    KEY_UP,
};
enum LedState {
```

```c
        LED_ON = 0,
        LED_OFF,
        LED_SPARK,
};

enum LedState g_ledState = LED_SPARK;

static volatile WifiIotGpioValue value = WIFI_IOT_GPIO_VALUE0;

static void OnKeyDown(char *arg)
{
        (void)arg;
        //WifiIoTGpioSetIsrMask(KEY_GPIO, 1);
        value = !value;
        //WifiIoTGpioSetIsrMask(KEY_GPIO, 0);
}

static void *KeyIntTask(const char *arg)
{
        (void)arg;

        while (1) {

                if (!value){
                        switch (g_ledState) {
                                case LED_ON:
                                GpioSetOutputVal(LED_TEST_GPIO, 0);
                                usleep(LED_INTERVAL_TIME_US);
                                break;
                                case LED_OFF:
                                GpioSetOutputVal(LED_TEST_GPIO, 1);
                                usleep(LED_INTERVAL_TIME_US);
                                break;
                                case LED_SPARK:
                                GpioSetOutputVal(LED_TEST_GPIO, 0);
                                usleep(LED_INTERVAL_TIME_US);
                                GpioSetOutputVal(LED_TEST_GPIO, 1);
                                usleep(LED_INTERVAL_TIME_US);
                                break;
                                default:
                                usleep(LED_INTERVAL_TIME_US);
                                break;
                        }
                }

        }

        return NULL;
}

static void KeyIntEntry(void)
```

```c
{
    osThreadAttr_t attr;

    GpioInit(LED_TEST_GPIO);
    GpioSetDir(LED_TEST_GPIO, WIFI_IOT_GPIO_DIR_OUT);

    GpioInit(KEY_GPIO);
    GpioSetDir(KEY_GPIO, WIFI_IOT_GPIO_DIR_IN);
    IoSetFunc(WIFI_IOT_IO_NAME_GPIO_5,WIFI_IOT_IO_PULL_UP);
    GpioRegisterIsrFunc(KEY_GPIO,
    WIFI_IOT_INT_TYPE_EDGE,
    WIFI_IOT_CPIO_EDGE_FALL_LEVEL_LOW,
    OnKeyDown,NULL);

    attr.name = "KeyIntTask";
    attr.attr_bits = 0U;
    attr.cb_mem = NULL;
    attr.cb_size = 0U;
    attr.stack_mem = NULL;
    attr.stack_size = LED_TASK_STACK_SIZE;
    attr.priority = LED_TASK_PRIO;

    if (osThreadNew((osThreadFunc_t)KeyIntTask, NULL, &attr) == NULL) {
        printf("[InterruptExample] Falied to create KeyIntTask!\n");
    }
}

SYS_RUN(KeyIntEntry);
```

在包含示例程序 keyInt.c 的目录下,修改 BUILD.gn 文件,文件内容如下。

```
//定义名为 iotkey 的静态库
static_library("iotkey") {
    //编译该静态库需要的源代码文件列表
    sources = [
    "keyInt.c"
    ]
    //编译该静态库需要的包含目录列表
    include_dirs = [
    "//utils/native/lite/include",
    "//kernel/liteos_m/kal/cmsis",
    "//base/iot_hardware/peripheral/interfaces/kits",
    ]
}
```

在上级目录下,修改 applications/sample/wifi-iot/app/BUILD.gn 文件,使 keyInt.c 参与编译。

```
import("//build/lite/config/component/lite_component.gni")
    lite_component("app") {
        features = [
```

```
                "iotkey: iotkey"
            ]
        }
```

3. 实验结果

每按一次按键 S2,LED 状态改变一次。即如 LED 闪烁,按一次按键 S2 后,LED 熄灭;再按一次按键 S2,LED 闪烁。

3.5　Hi3861 PWM 输出实验

1. 实验内容

本实验是在 Hi3861 核心板和环境监测板上,完成通过 GPIO 实现按键中断事件对蜂鸣器的控制功能。

2. 实验代码

在 OpenHarmony 3.0 LTS 版本中,在\applications\sample\wifi-iot\app 目录下的 iothardware 目录中,新建程序代码文件 pwmout.c,该文件完整代码如下。

```c
#include <stdio.h>

#include <unistd.h>
#include "ohos_init.h"
#include "cmsis_os2.h"
#include "iot_gpio.h"
#include "iot_gpio_ex.h"
#include "iot_pwm.h"
#include "iot_watchdog.h"

#define PWM_TASK_STACK_SIZE 1024
#define PWM_TASK_PRIO 25
#define KEY_GPIO 5
#define PWM_IO 9

static int beepon = 1;

static void OnKeyDown(char * arg)
{
    (void)arg;
    //IoTGpioSetIsrMask(KEY_GPIO, 1);
    beepon = !beepon;
    //IoTGpioSetIsrMask(KEY_GPIO, 0);
    printf("Pressed Key! beepon = %d\n", beepon);
}

static void * PWMTask(const char * arg)
{
    (void)arg;
```

```
        printf("\n PWMTask is running!\n");
        while (1) {
            if (beepon){
                IoTPwmStart(0, 50, 40000);
                //IoTPwmStart(0, 40000000, 80000000);
                //IoTPwmStart(0, 40000, 160000);
                //unsleep(10000000);
            }else{
                IoTPwmStop(0);
            }

        }
        return NULL;
}

static void PWMTaskEntry(void)
{
        osThreadAttr_t attr;

        IoTGpioInit(PWM_IO);
        IoSetFunc(PWM_IO, IOT_IO_FUNC_GPIO_9_PWM0_OUT);
        IoTGpioSetDir(PWM_IO, IOT_GPIO_DIR_OUT);
        IoTPwmInit(0);

        IoTGpioDeinit(KEY_GPIO);
        IoTGpioInit(KEY_GPIO);
        IoSetFunc(IOT_IO_NAME_GPIO_5, IOT_IO_FUNC_GPIO_5_GPIO);
        IoTGpioSetDir(KEY_GPIO, IOT_GPIO_DIR_IN);
        IoSetPull(KEY_GPIO,1);
        IoTGpioRegisterIsrFunc(KEY_GPIO,
        IOT_INT_TYPE_EDGE,
        IOT_GPIO_EDGE_FALL_LEVEL_LOW,
        OnKeyDown,NULL);
        IoTWatchDogDisable();

        attr.name = "PWMTask";
        attr.attr_bits = 0U;
        attr.cb_mem = NULL;
        attr.cb_size = 0U;
        attr.stack_mem = NULL;
        attr.stack_size = PWM_TASK_STACK_SIZE;
        attr.priority = PWM_TASK_PRIO;

        if (osThreadNew((osThreadFunc_t)PWMTask, NULL, &attr) == NULL) {
                printf("[PWMExample] Falied to create PWMTask!\n");
        }
}

SYS_RUN(PWMTaskEntry);
```

在包含程序文件 pwmout.c 的目录下,修改 BUILD.gn 文件,文件内容如下。

```
        static_library("ledexample") {
                    sources = [
                    #"led_example.c",
                    "pwmout.c"
                    ]

                    include_dirs = [
                    "//utils/native/lite/include",
                    "//kernel/liteos_m/kal/cmsis",
                    "//base/iot_hardware/peripheral/interfaces/kits",
                    ]
        }
```

在上级目录下,修改 applications/sample/wifi-iot/app/BUILD.gn 文件,引入 ledexample 库。

```
import("//build/lite/config/component/lite_component.gni")

lite_component("app") {
        features = [
        #"startup"
        "iothardware:ledexample"
        #"testapp:myapp",
        #"iotkey:iotkey"
        ]
}
```

编译之前,需要修改一个配置文件,该文件为 LiteOS 的用户配置文件 usr_config.mk,存放在目录/device/hisilicon/hispark_pegasus/sdk_liteos/build/config 下。在目录/device/hisilicon/hispark_pegasus/sdk_liteos/tools/menuconfig 中,也有配置文件 usr_config.mk,该文件用于 menuconfig 工具,在开发软件 LiteOS Studio 中可以通过图形界面配置。

在配置文件 usr_config.mk 中,找到如下配置。

```
# BSP Settings
#
# CONFIG_I2C_SUPPORT is not set
# CONFIG_I2S_SUPPORT is not set
# CONFIG_SPI_SUPPORT is not set
# CONFIG_DMA_SUPPORT is not set
# CONFIG_SDIO_SUPPORT is not set
# CONFIG_SPI_DMA_SUPPORT is not set
# CONFIG_UART_DMA_SUPPORT is not set
# CONFIG_PWM_SUPPORT is not set
# CONFIG_PWM_HOLD_AFTER_REBOOT is not set
CONFIG_AT_SUPPORT = y
# CONFIG_FILE_SYSTEM_SUPPORT is not set
CONFIG_UART0_SUPPORT = y
CONFIG_UART1_SUPPORT = y
# CONFIG_UART2_SUPPORT is not set
```

修改其中关于 PWM 的配置项,如下。

```
# BSP Settings
#
# CONFIG_I2C_SUPPORT is not set
# CONFIG_I2S_SUPPORT is not set
# CONFIG_SPI_SUPPORT is not set
# CONFIG_DMA_SUPPORT is not set
# CONFIG_SDIO_SUPPORT is not set
# CONFIG_SPI_DMA_SUPPORT is not set
# CONFIG_UART_DMA_SUPPORT is not set
CONFIG_PWM_SUPPORT = y
CONFIG_PWM_HOLD_AFTER_REBOOT = y
CONFIG_AT_SUPPORT = y
# CONFIG_FILE_SYSTEM_SUPPORT is not set
CONFIG_UART0_SUPPORT = y
CONFIG_UART1_SUPPORT = y
# CONFIG_UART2_SUPPORT is not set
# end of BSP Settings
```

3. 实验结果

烧录完成后。由于 GPIO9 端口同时通过 J3 连接 LED 灯,该实验运行时要去除核心板上的 J3 跳线帽。复位 Hi3861 核心板,蜂鸣器发出声音。

在串口调试区,可以看到如下输出。

```
ready to OS start
sdk ver: Hi3861V100R001C00SPC025 2020-09-03 18:10:00
FileSystem mount ok.
wifi init success!
hilog will init.
hievent will init.
hievent init success.

PWMTask is running!
No crash dump found!
```

按下按键 S2,则输出信息如下。

```
Pressed Key! beepon = 0
Pressed Key! beepon = 1
Pressed Key! beepon = 0
Pressed Key! beepon = 1
Pressed Key! beepon = 0
```

如没有环境监测板,也可通过连接跳线帽 J3 后,通过板载 LED 灯观察实验现象。输出 PWM 时,beepon = 1,板载 LED 灯亮度会变暗。未输出 PWM 时,beepon = 0,板载 LED 灯亮度较亮。

4. 扩展

PWM 波形是固定频率(或固定周期)的信号,占空比可变的方波。占空比是一个周期内高电平时间和低电平时间的比例,一个周期内高电平时间长,占空比大,反之占空比小。占空比用百分比表示,范围为 0%~100%。

PWM 通过数字方式控制模拟电路,大幅度降低系统的成本和功耗。通过改变 PWM 脉冲的周期可以调频,改变 PWM 脉冲的宽度或占空比可以调压,采用适当控制方法即可使电压与频率协调变化。可以通过调整 PWM 的周期、PWM 的占空比而达到控制充电电流的目的。

PWM 广泛应用在很多领域,包括测量、通信、电机控制、伺服控制、调光、开关电源、功率控制与变换等领域。如航模中的控制信号大多是 PWM 信号,工业上 PID 控制的温控信号可以使用 PWM 脉冲,PWM 控制技术在逆变电路中也应用广泛。

3.6 Hi3861 I2C 读取 AHT 实验

1. 实验内容

本实验是在 Hi3861 核心板和环境监测板上,实现通过 GPIO 读取 AHT20 温湿度传感器的功能。

2. 实验代码

在 OpenHarmony 3.0 LTS 版本中,在 /applications/sample/wifi-iot/app 目录下的 aht 目录中,新建程序代码文件 aht20.c,该文件完整代码如下。

```
#include <stdio.h>

#include <unistd.h>

#include "ohos_init.h"
#include "cmsis_os2.h"
#include "iot_errno.h"

#include "iot_gpio.h"
#include "iot_gpio_ex.h"
#include "iot_i2c.h"

#define AHT_TIME_STARTUP            40 * 1000//40ms
#define AHT_TIME_CALIBRATION        (10 * 1000) //10ms
#define AHT_TIME_MEASURE            (80 * 1000) //80ms

#define AHT_CMD_CALIBRATION         (0xBE) //aht init cmd
#define AHT_CMD_CALIBRATION_0       (0x08)
#define AHT_CMD_CALIBRATION_1       (0x00)

#define AHT_CMD_TRIGGER             (0xAC) // measure cmd
#define AHT_CMD_TRIGGER_0           (0x33)
#define AHT_CMD_TRIGGER_1           (0x00)
```

```c
#define AHT_CMD_RESET              (0xBA)

#define AHT_CMD_GETSTATUS          (0x71)

#define BAUDRATE_INIT              (400000)
#define IotI2cIdx                   0

#define AHT_DEVICE_ADDR            (0x38) //7b 设备地址
#define AHT_READ_ADDR              (0x38 << 1)|0x1 //读设备地址,7b 设备地址 + 1
#define AHT_WRITE_ADDR             (0x38 << 1)|0x0 //写设备地址,7b 设备地址 + 0

#define     AHT_REG_ARRAY_LEN       (6)
#define     AHT_REG_ARRAY_LEN_INIT  1 //0x71 获取状态后的状态字节

#define AHT_STATUS_BUSY_SHIFT      7 //bit[7]为忙闲指示位
#define AHT_STATUS_BUSY_MASK       (0x1 << AHT_STATUS_BUSY_SHIFT)//0b1000'0000
#define AHT_STATUS_BUSY(status)    ((status & AHT_STATUS_BUSY_MASK) >> AHT_STATUS_BUSY_SHIFT)

#define AHT_STATUS_CALI_SHIFT      3 //bit[3]为校准指示位
#define AHT_STATUS_CALI_MASK       (0x1 << AHT_STATUS_CALI_SHIFT)//0b0000'1000
#define AHT_STATUS_CALI(status)    ((status & AHT_STATUS_CALI_MASK)>> AHT_STATUS_CALI_SHIFT)

#define AHT_RETRY_MAXTIME          3
typedef struct {
    /** 发送数据缓冲区指针 */
    unsigned char * sendBuf;
    /** 发送数据长度 */
    unsigned int   sendLen;
    /** 接收数据缓冲区指针 */
    unsigned char * receiveBuf;
    /** 接收数据长度 */
    unsigned int   receiveLen;
} IotI2cData;

/*
 * 读 AHT 到 buffer
 */
static uint32_t AHT_Read ( uint8_t * buffer, uint32_t bufflen)
{
    IotI2cData data = {0};
    data.receiveBuf = buffer;
    data.receiveLen = bufflen;
    uint32_t ret = IoTI2cRead(IotI2cIdx, AHT_READ_ADDR, &data, sizeof(data));
    if (ret != IOT_SUCCESS){
        printf("IoTI2cRead() failed, %0X!\n", ret);
        return ret;
    }
    return IOT_SUCCESS;
```

```c
}

/*
 * 读 AHT 到 buffer
 */
static uint32_t AHT_Write ( uint8_t* buffer, uint32_t bufflen)
{
    IotI2cData data = {0};
    data.sendBuf = buffer;
    data.sendLen = bufflen;
    uint32_t ret = IoTI2cWrite(IotI2cIdx, AHT_WRITE_ADDR, &data, sizeof(data));
    if (ret != IOT_SUCCESS){
        printf("IoTI2cWrite(%02X) failed, %0X!\n", buffer[0], ret);
        return ret;
    }
    return IOT_SUCCESS;
}

/*
 * 向 AHT20 发送 0x71 命令,获取状态
 */
static uint32_t AHT_StatusCommand(void)
{
    uint8_t statusCmd[] = {AHT_CMD_GETSTATUS};
    return AHT_Write(statusCmd, sizeof(statusCmd));
}

/*
 * 向 AHT20 发送 0xBA 命令,软复位
 */
static uint32_t AHT_ResetCommand(void)
{
    uint8_t resetCmd[] = {AHT_CMD_RESET};
    return AHT_Write(resetCmd, sizeof(resetCmd));
}

/*
 * 向 AHT20 发送 0xBE0800 命令,初始化校准
 */
static uint32_t AHT_CalibrateCommand(void)
{
    uint8_t calibrateCmd[] = {AHT_CMD_CALIBRATION,
             AHT_CMD_CALIBRATION_0,AHT_CMD_CALIBRATION_1};
    return AHT_Write(calibrateCmd, sizeof(calibrateCmd));
}

/*
 * 向 AHT20 发送 0xAC3300 命令,开始测量
 */
static uint32_t AHT_StartMeasure(void)
{
```

```c
        uint8_t MeaCmd[] = {AHT_CMD_TRIGGER,
                AHT_CMD_TRIGGER_0, AHT_CMD_TRIGGER_1};
        return AHT_Write(MeaCmd, sizeof(MeaCmd));
}

/*
 * 校准数据
 */
uint32_t AHT_Calibrate(void)
{
        uint32_t ret = 0;
        uint8_t buffer[AHT_REG_ARRAY_LEN_INIT] = {AHT_CMD_GETSTATUS};
        memset(&buffer, 0x0, sizeof(buffer));

        ret = AHT_StatusCommand();
        if (ret != IOT_SUCCESS)
        return ret;

        ret = AHT_Read(buffer, sizeof(buffer));
        if (ret != IOT_SUCCESS)
        return ret;

        if (AHT_STATUS_BUSY(buffer[0]) || !AHT_STATUS_CALI(buffer[0])) {
                ret = AHT_ResetCommand();
                if (ret != IOT_SUCCESS)
                return ret;
                usleep(AHT_TIME_STARTUP);
                ret = AHT_CalibrateCommand();
                usleep(AHT_TIME_CALIBRATION);
                return ret;
        }
        return IOT_SUCCESS;

}


uint32_t AHT_ObtainRlt(float * tempVal, float * humiVal)
{
        uint32_t ret = 0;
        uint32_t i = 0;
        if (tempVal == NULL || humiVal == NULL){
                return IOT_FAILURE;
        }

        uint8_t buffer[AHT_REG_ARRAY_LEN] = {0};
        memset(&buffer, 0x0, sizeof(buffer));

        ret = AHT_Read(buffer, sizeof(buffer));
        if(ret != IOT_SUCCESS){
                return ret;
```

```c
        }

        for(i = 0; AHT_STATUS_BUSY(buffer[0]) && i < AHT_RETRY_MAXTIME; i++){
                usleep(AHT_TIME_MEASURE);
                ret = AHT_Read(buffer, sizeof(buffer));
                if(ret != IOT_SUCCESS){
                        return ret;
                }
        }

        if( i >= AHT_RETRY_MAXTIME){
                printf("AHT always busy!\r\n");
                return IOT_FAILURE;
        }

        uint32_t humiOringi = buffer[1];
        humiOringi = (humiOringi << 8) | buffer[2];
        humiOringi = (humiOringi << 4) | ((buffer[3] &0xf0)>> 4);
        * humiVal = humiOringi/pow(2,20) * 100;

        uint32_t tempOringi = buffer[3]&0x0f;
        tempOringi = (tempOringi << 8) | buffer[4];
        tempOringi = (tempOringi << 8) | buffer[5];
        * tempVal = tempOringi/pow(2,20) * 200 - 50;

        return IOT_SUCCESS;
}

static void TempHumiTask(void * arg)
{
        (void) arg;
        uint32_t ret = 0;
        float humirslt = 0.0f;
        float temprslt = 0.0f;


        while( AHT_Calibrate()!= IOT_SUCCESS){
                printf("AHT sensor init failed! \r\n");
                usleep(1000);
        }

        while(1){
                ret = AHT_StartMeasure();
                if (ret != IOT_SUCCESS){
                        printf("TAHT trigger measure failed! \r\n");
                }
                ret = AHT_ObtainRlt(&temprslt, &humirslt);
                if(ret != IOT_SUCCESS){
                        printf("Failed to obtain result!\r\n");
                }
                printf("temprature: %.2f\n",temprslt);
```

```c
                printf("humidity: %.2f\n",humirslt);
                sleep(1);
        }
}

static void StartAhtTaskEntry(void)
{
        osThreadAttr_t attr;

        IoTGpioInit(13);
        IoSetFunc(13, 6);
        IoTGpioInit(14);
        IoSetFunc(14, 6);
        IoTI2cInit(IotI2cIdx, BAUDRATE_INIT);

        attr.stack_size = 4096;
        attr.priority = 25;
        attr.name = "app_demo_aht20_task";

        if (osThreadNew((osThreadFunc_t)TempHumiTask, NULL, &attr) == NULL) {
                printf("[AHTExample] Falied to create TempHumiTask!\n");
        }
}

SYS_RUN(StartAhtTaskEntry);
```

在包含程序文件 aht20.c 的目录下，新建或修改 BUILD.gn 文件，文件内容如下。

```
static_library("app_demo_AHT") {
        sources = [
        "aht20.c"
        ]

        include_dirs = [
        "./",
        "//utils/native/lite/include",
        "//kernel/liteos_m/kal/cmsis",
        "//base/iot_hardware/peripheral/interfaces/kits",
        ]
}
```

在上级目录下，修改 applications/sample/wifi-iot/app/BUILD.gn 文件，引入 app_demo_AHT 库。

```
mport("//build/lite/config/component/lite_component.gni")

lite_component("app") {
        features = [
                "aht: app_demo_AHT"
        ]
}
```

编译之前，需要修改一个配置文件，该文件为 LiteOS 的用户配置文件 usr_config.mk，存放在目录/device/hisilicon/hispark_pegasus/sdk_liteos/build/config 下。在配置文件 usr_config.mk 中，找到如下配置。

```
# BSP Settings
#
# CONFIG_I2C_SUPPORT is not set
# CONFIG_I2S_SUPPORT is not set
# CONFIG_SPI_SUPPORT is not set
```

修改其中关于 I2C 的配置项，如下。

```
# BSP Settings
#
# CONFIG_I2C_SUPPORT is not set
CONFIG_I2C_SUPPORT = y
# CONFIG_I2S_SUPPORT is not set
# CONFIG_SPI_SUPPORT is not set
```

3. 实验结果

烧录完成后，复位 Hi3861 核心板，在串口调试区，可以看到如下输出。

```
ready to OS start
sdk ver:Hi3861V100R001C00SPC025 2020-09-03 18:10:00
FileSystem mount ok.
wifi init success!
hilog will init.
hievent will init.
hievent init success.

temprature: 16.42
humidity: 40.61
```

4. 扩展

I2C(inter-integrated circuit)总线是由 Philips 公司开发的两线式串行总线，由数据线 SDA 和时钟线 SCL 构成，数据线用来传输数据，时钟线用来同步数据收发。SDA 和 SCL 是双向的，通过一个电流源或上拉电阻连接到正电压。I2C 总线上每个器件有一个唯一的地址，同时支持多个主机和多个从机，连接到总线的接口数量由总线电容限制决定。标准模式、快速模式和高速模式下数据传输速率分别为 100kb/s、400kb/s 和 3.4Mb/s。

I2C 总线空闲状态时，SDA 和 SCL 信号同时处于高电平，此时各个器件的输出级场效管均处在截止状态，即释放总线，由两条信号线各自的上拉电阻把电平拉高。在 SCL 高电平周期内，SDA 线电平必须保持稳定，SDA 线仅可以在 SCL 为低电平时改变。

I2C 通信过程由起始、结束、发送、应答和接收五个部分构成。

3.7 Hi3861 AT 指令实验

1. 实验内容

本实验通过 AT 指令控制 Hi3861 连接 WiFi 热点，并与指定 IP 地址进行连通性测试。

2. 实验过程与结果

实验过程结果如下。

```
AT + IFCFG
+ IFCFG: lo,ip = 127.0.0.1,netmask = 255.0.0.0,gateway = 127.0.0.1,
ip6 = : : 1,HWaddr = 00,MTU = 16436,LinkStatus = 1,RunStatus = 1
OK

AT + SCAN
ERROR

AT + SCANSSID
ERROR

AT + STARTSTA
OK

AT + SCAN
OK

+ NOTICE: SCANFINISH
AT + SCANRESULT
+ SCANRESULT: tl,80: d0: 9b: 69: 64: e3,6, - 45,2
+ SCANRESULT: ChinaNet - ****,00: 06: 0c: e0: 10: f1,1, - 61,3
+ SCANRESULT: ChinaNet - ****,54: 00: 00: 46: 01: dd,8, - 82,3
+ SCANRESULT: ChinaNet - ****,00: 05: 11: 35: 00: 4c,6, - 83,3
+ SCANRESULT: ,10: 32: 06: 07: c2: 81,6, - 66,0
OK

AT + CONN = "tl",,3,"11111111"
+ NOTICE: DISCONNECTED
OK

+ NOTICE: SCANFINISH
+ NOTICE: CONNECTED
AT + IFCFG
+ IFCFG: wlan0,ip = 0.0.0.0,netmask = 0.0.0.0,gateway = 0.0.0.0,ip6 =
FE80: : B6C9: B9FF: FEAF: 6661,ip6 = 240E: 45E: C41: 2926: B6C9: B9FF: FEAF: 6661,
HWaddr = b4: c9: b9: af: 66: 61,MTU = 1500,LinkStatus = 1,RunStatus = 1
+ IFCFG: lo,ip = 127.0.0.1,netmask = 255.0.0.0,gateway = 127.0.0.1,
ip6 = : : 1,HWaddr = 00,MTU = 16436,LinkStatus = 1,RunStatus = 1
OK

AT + PING = www.huawei.com
ping : Host : www.huawei.com can't be resolved to IP address
ERROR

AT + DHCP = wlan0,1
OK
```

```
AT + IFCFG
 + IFCFG: wlan0,ip = 192.168.43.78,netmask = 255.255.255.0,gateway = 192.168.43.1,
ip6 = FE80:: B6C9: B9FF: FEAF: 6661,ip6 = 240E: 45E: C41: 2926: B6C9: B9FF: FEAF: 6661,
HWaddr = b4: c9: b9: af: 66: 61,MTU = 1500,LinkStatus = 1,RunStatus = 1
 + IFCFG: lo,ip = 127.0.0.1,netmask = 255.0.0.0,gateway = 127.0.0.1,ip6 = :: 1,
HWaddr = 00,MTU = 16436,LinkStatus = 1,RunStatus = 1
OK

AT + PING = www.gov.cn
 + PING:

[0]Reply from 218.77.92.141: time = 43ms TTL = 52
[1]Reply from 218.77.92.141: time = 43ms TTL = 52
[2]Reply from 218.77.92.141: time = 44ms TTL = 52
[3]Reply from 218.77.92.141: time = 28ms TTL = 52
4 packets transmitted, 4 received, 0 loss, rtt min/avg/max = 28/39/44 ms

OK
```

3. 扩展

Hi3861 核心板提供了丰富的 AT 指令，用于 PC 等用户终端和 Hi3861 之间控制信息的交互。

AT	AT + RST	AT + MAC	AT + HELP
AT + SYSINFO	AT + DHCP	AT + DHCPS	AT + NETSTAT
AT + PING	AT + PING6	AT + DNS	AT + DUMP
AT + IPSTART	AT + IPLISTEN	AT + IPSEND	AT + IPCLOSE
AT + XTALCOM	AT + RDTEMP	AT + STOPSTA	AT + SCAN
AT + SCANCHN	AT + SCANSSID	AT + SCANPRSSID	AT + SCANRESULT
AT + CONN	AT + FCONN	AT + DISCONN	AT + STASTAT
AT + RECONN	AT + STARTAP	AT + SETAPADV	AT + STOPAP
AT + SHOWSTA	AT + DEAUTHSTA	AT + APSCAN	AT + RXINFO
AT + CC	AT + TPC	AT + TRC	AT + SETRATE
AT + ARLOG	AT + VAPINFO	AT + USRINFO	AT + SLP
AT + WKGPIO	AT + USLP	AT + ARP	AT + PS
AT + ND	AT + CSV	AT + FTM	AT + FTMERASE
AT + SETUART	AT + IFCFG	AT + STARTSTA	AT + ALTX
AT + ALRX	AT + CALFREQ	AT + SETRPWR	AT + RCALDATA
AT + SETIOMODE	AT + GETIOMODE	AT + GPIODIR	AT + WTGPIO
AT + RDGPIO	AT + SYSPARA	AT + HIDUMPER	

AT 指令包括测试指令，格式为"AT+<cmd>=?"；查询指令，格式为"AT+<cmd>?"；设置指令，格式为"AT+<cmd>=<parameter>,"；执行指令，格式为"AT+<cmd>"。

在 DevEco Device Tool 的工程任务中，单击 Monitor 进入串口调试状态，如图 3.2 所示，在终端输入 AT 命令。按 Ctrl+C 组合键退出该状态。

Hi3861 和 WiFi 相关的常用 AT 指令有如下几个。

（1）AT 用于测试 AT 功能。

（2）AT+NETSTAT 用于查看网络状态。

图 3.2 AT 指令测试

(3) AT+RST 用于复位命令。

(4) AT+STARTSTA 用于启动 STA。

(5) AT+STOPSTA 用于关闭 STA。

(6) AT+SCAN 用于启动 STA 扫描。

(7) AT+SCANRESULT 用于查看 STA 扫描结果。

(8) AT+CONN=<ssid>,<bssid>,<auth_type>[,<passwd>]用于发起与 AP 的连接。

(9) AT+DHCP=wlan0,1 dhcp 为客户端命令。

(10) AT+IFCFG 为网络接口命令。

(11) AT+PING=目的 IP 为地址 ping 命令。

(12) AT+DISCONN 为关闭连接。

3.8 Hi3861 WiFi 连接实验

1. 实验内容

本实验中,Hi3861 以 WiFi 的 station 模式工作,连接到 AP,AP 名为 tl。

2. 实验代码

在 OpenHarmony 3.0 LTS 版本中,在/applications/sample/wifi-iot/app 目录下的 wificon 目录中,新建头文件 wifista.h 和程序代码文件 wifista.c。头文件 wifista.h 完整代码如下:

```c
#define PARAM_HOTSPOT_SSID "tl"            //AP 的 SSID
#define PARAM_HOTSPOT_PSK  "testtest"      //AP 的 PSK
#define PARAM_HOTSPOT_TYPE WIFI_SEC_TYPE_PSK //在 wifi_device_config.h 中定义
//#define PARAM_SERVER_ADDR "192.168.43.165"
//计算机的 IP 地址
#define PARAM_SERVER_PORT 5678
```

源代码文件 wifista.c 完整代码如下。

```c
#include <stdio.h>
#include <string.h>
#include <unistd.h>

#include "ohos_init.h"
#include "cmsis_os2.h"

#include "wifi_sta.h"

#include "wifi_device.h"
#include "lwip/netifapi.h"
#include "lwip/api_shell.h"

static void PrintLinkedInfo(WifiLinkedInfo* info)
{
    if (!info) return;

    static char macAddress[32] = {0};
    unsigned char* mac = info->bssid;
    snprintf(macAddress, sizeof(macAddress), "%02X:%02X:%02X:%02X:%02X:%02X",
        mac[0], mac[1], mac[2], mac[3], mac[4], mac[5]);
    printf("bssid: %s, rssi: %d, connState: %d, reason: %d, ssid: %s\r\n",
        macAddress, info->rssi, info->connState, info->disconnectedReason, info->ssid);
}

static volatile int g_connected = 0;

static void OnWifiConnectionChanged(int state, WifiLinkedInfo* info)
{
    if (!info) return;

    printf("%s %d, state = %d, info = \r\n", __FUNCTION__, __LINE__, state);
    PrintLinkedInfo(info);

    if (state == WIFI_STATE_AVALIABLE) {
        g_connected = 1;
    } else {
        g_connected = 0;
    }
}

static void OnWifiScanStateChanged(int state, int size)
```

```c
{
    printf("%s %d, state = %X, size = %d\r\n", __FUNCTION__, __LINE__, state, size);
}

static WifiEvent g_defaultWifiEventListener = {
    .OnWifiConnectionChanged = OnWifiConnectionChanged,
    .OnWifiScanStateChanged = OnWifiScanStateChanged
};

static struct netif * g_iface = NULL;

int ConnectToHotspot(WifiDeviceConfig * apConfig)
{
    WifiErrorCode errCode;
    int netId = -1;

    errCode = RegisterWifiEvent(&g_defaultWifiEventListener);
    printf("RegisterWifiEvent: %d\r\n", errCode);

    errCode = EnableWifi();
    printf("EnableWifi: %d\r\n", errCode);

    errCode = AddDeviceConfig(apConfig, &netId);
    printf("AddDeviceConfig: %d\r\n", errCode);

    g_connected = 0;
    errCode = ConnectTo(netId);
    printf("ConnectTo(%d): %d\r\n", netId, errCode);

    while (!g_connected) {              //等待直到连接至 AP
        osDelay(10);
    }
    printf("g_connected: %d\r\n", g_connected);

    g_iface = netifapi_netif_find("wlan0");
    if (g_iface) {
        err_t ret = 0;
        char * hostname = "EnvironSta01";
        ret = netifapi_set_hostname(g_iface, hostname, strlen(hostname));
        printf("netifapi_set_hostname: %d\r\n", ret);

        ret = netifapi_dhcp_start(g_iface);
        printf("netifapi_dhcp_start: %d\r\n", ret);

        osDelay(100);                   //等待 DHCP 服务器返回 IP
        #if 1
        ret = netifapi_netif_common(g_iface, dhcp_clients_info_show, NULL);
        printf("netifapi_netif_common: %d\r\n", ret);
        #else
        //下面这种方式也可以打印 IP 地址、网关、子网掩码信息
```

```c
                        ip4_addr_t ip = {0};
                        ip4_addr_t netmask = {0};
                        ip4_addr_t gw = {0};
                        ret = netifapi_netif_get_addr(g_iface, &ip, &netmask, &gw);
                        if (ret == ERR_OK) {
                                printf("ip = %s\r\n", ip4addr_ntoa(&ip));
                                printf("netmask = %s\r\n", ip4addr_ntoa(&netmask));
                                printf("gw = %s\r\n", ip4addr_ntoa(&gw));
                        }
                        printf("netifapi_netif_get_addr: %d\r\n", ret);
                        #endif
                }
        return netId;
}

void DisconnectWithHotspot(int netId)
{
        if (g_iface) {
                err_t ret = netifapi_dhcp_stop(g_iface);
                printf("netifapi_dhcp_stop: %d\r\n", ret);
        }

        WifiErrorCode errCode = Disconnect();              //与AP断开连接
        printf("Disconnect: %d\r\n", errCode);

        errCode = UnRegisterWifiEvent(&g_defaultWifiEventListener);
        printf("UnRegisterWifiEvent: %d\r\n", errCode);

        RemoveDevice(netId);                               //清除AP配置
        printf("RemoveDevice: %d\r\n", errCode);

        errCode = DisableWifi();
        printf("DisableWifi: %d\r\n", errCode);
}

static void NetDemoTask(void *arg)
{
        (void)arg;
        WifiDeviceConfig config = {0};

        //准备AP的配置参数
        strcpy(config.ssid, PARAM_HOTSPOT_SSID);
        strcpy(config.preSharedKey, PARAM_HOTSPOT_PSK);
        config.securityType = PARAM_HOTSPOT_TYPE;

        osDelay(10);

        int netId = ConnectToHotspot(&config);

        int timeout = 10;
        while (timeout--) {
```

```
                printf("After %d seconds, I will exit!\r\n", timeout);
                osDelay(100);
        }

        printf("disconnect to AP ...\r\n");
        DisconnectWithHotspot(netId);
        printf("disconnect to AP done!\r\n");
}

static void NetDemoEntry(void)
{
        osThreadAttr_t attr;

        attr.name = "NetDemoTask";
        attr.attr_bits = 0U;
        attr.cb_mem = NULL;
        attr.cb_size = 0U;
        attr.stack_mem = NULL;
        attr.stack_size = 10240;
        attr.priority = osPriorityNormal;

        if (osThreadNew(NetDemoTask, NULL, &attr) == NULL) {
                printf("[NetDemoEntry] Falied to create NetDemoTask!\n");
        }
}

SYS_RUN(NetDemoEntry);
```

在包含程序文件 wifista.c 的目录下，新建或修改 BUILD.gn 文件，文件内容如下。

```
static_library("net_demo") {
        sources = ["wifista.c"]

        include_dirs = [
        "//utils/native/lite/include",
        "//kernel/liteos_m/components/cmsis/2.0",
        "//base/iot_hardware/interfaces/kits/wifiiot_lite",
        "//base/iot_hardware/peripheral/interfaces/kits",
        "//foundation/communication/wifi_lite/interfaces/wifiservice"
        ]
}
```

在上级目录下，修改 applications/sample/wifi-iot/app/BUILD.gn 文件，引入 app_demo_wifi 库。

```
import("//build/lite/config/component/lite_component.gni")

lite_component("app") {
        features = [
        "wificon: net_demo"
        ]
```

}

编译之前,需要修改 LiteOS 的用户配置文件 usr_config.mk,存放在目录/device/hisilicon/hispark_pegasus/sdk_liteos/build/config 下。在配置文件 usr_config.mk 中,找到如下配置。

```
# Lwip Settings
#
CONFIG_DHCPS_GW = y
# CONFIG_NETIF_HOSTNAME is not set
# CONFIG_DHCP_VENDOR_CLASS_IDENTIFIER is not set
# CONFIG_LWIP_LOWPOWER is not set
# end of Lwip Settings
```

修改其中关于 lwip 的配置项,如下。

```
# Lwip Settings
#
CONFIG_DHCPS_GW = y
# CONFIG_NETIF_HOSTNAME is not set
CONFIG_NETIF_HOSTNAME = y
# CONFIG_DHCP_VENDOR_CLASS_IDENTIFIER is not set
# CONFIG_LWIP_LOWPOWER is not set
# end of Lwip Settings
```

3. 实验结果

烧录完成后,复位 Hi3861 核心板,在串口调试区,可以看到如下输出。

```
ready to OS start
sdk ver: Hi3861V100R001C00SPC025 2020-09-03 18:10:00
formatting spiffs...
FileSystem mount ok.
wifi init success!
hilog will init.
hievent will init.
hievent init success.
hiview init success.RegisterWifiEvent: 0
EnableWifi: 0
AddDeviceConfig: 0
ConnectTo(1): 0
No crash dump found!
+NOTICE: SCANFINISH
+NOTICE: CONNECTED
OnWifiConnectionChanged 54, state = 1, info =
bssid: 80:D0:9B:69:64:E3, rssi: 0, connState: 0, reason: 0, ssid: tl
g_connected: 1
netifapi_set_hostname: 0
netifapi_dhcp_start: 0
server :
server_id : 192.168.43.1
mask : 255.255.255.0, 1
```

```
gw: 192.168.43.1
T0: 3600
T1: 1800
T2: 3150
clients <1>:
mac_idx mac              addr            state   lease   tries   rto
0       b4c9b9af6661     192.168.43.78   10      0       1       4
netifapi_netif_common: 0
After 9 seconds, I will exit!
...
After 0 seconds, I will exit!
disconnect to AP
netifapi_dhcp_stop: 0
+ NOTICE: DISCONNECTED
OnWifiConnectionChanged 54, state = 0, info =
bssid: 80: D0: 9B: 69: 64: E3, rssi: 0, connState: 0, reason: 3, ssid:
Disconnect: 0
UnRegisterWifiEvent: 0
RemoveDevice: 0
DisableWifi: 0
disconnect to AP done!
```

在接入点(此处接入点名称为 tl)可以看到主机名 EnvironSta01 连接成功,如图 3.3 所示,并在 10s 后断开连接。

图 3.3 AP 显示已连接

4. 扩展

WiFi 的主要术语有接入点、工作站和 SSID 等。

接入点(Access Point,AP)是无线网络中的特殊节点,通过这个节点,无线网络中的其他类型节点可以和无线网络外部以及内部进行通信。

工作站(Station)表示通过 AP 连接到无线网络中的设备,这些设备可以和内部其他设备或者无线网络外部通信。

关联(Associate),指如果一个 Station 想要加入到无线网络中,需要和这个无线网络中的 AP 关联。

服务集标识(Service Set IDentifier,SSID)用来标识一个无线网络,可以将一个无线局域网分为几个需要不同身份验证的子网络,每一个子网络都需要独立的身份验证,只有通过身份验证的用户才可以进入相应的子网络,防止未被授权的用户进入本网络。

基本服务集标志符(BSSID)用来标识一个 BSS,其格式和 MAC 地址一样,是 48 位的地址格式。一般来说,它就是所处的无线接入点的 MAC 地址。

基本服务集(Basic Service Set,BSS)由一组相互通信的工作站组成。主要有两种,独立的基本服务组合称为 IBSS(Independent BSS),或者 Ad Hoc BSS,以及基础型基本服务组合 Infrastructure BSS。

接收信号强度(Received Signal Strength Indication,RSSI),通过 STA 扫描到 AP 站点的信号强度。

连接 WiFi 过程包含三个过程:扫描网络(Scanning)、认证过程(Authentication)和关联过程(Association)。

(1)扫描网络阶段。Station 通过主动扫描和被动扫描两种方式获取无线网络信息。主动扫描时,Station 定期发送 Probe Request 帧。如果扫描指定 SSID,如果有就该 AP 返回 Probe Response。如果扫描时发送广播 Probe Request,Station 会定期在其支持的信道列表中,发送 Probe Request 扫描网络,AP 会回复 Probe Response,Station 就会显示所有的 SSID 信息。被动扫描时,AP 每隔一段时间发送 Beacon 信标帧,提供 AP 和 BSS 相关信息,Station 会接收,显示可以加入的网络。

(2)认证过程可以通过 AP 和 Station 配置相同的共享密钥的共享密钥认证进行,也可以开放系统认证进行。

(3)关联过程中,Station 将速率、支持的信道,支持 QoS 的能力,以及选择的认证和加密算法,发送给 AP,AP 认证通过后返回一个唯一识别码,告诉 Station 认证成功。

3.9 Hi3861 MQTT 客户端实验

1. 实验内容

本实验的功能是通过 MQTT 协议发送消息给服务器以及接收订阅的消息。

2. 实验代码

在 OpenHarmony 3.0 LTS 版本中,在/applications/sample/wifi-iot/app 目录下的 mqtt_test 目录中,新建头文件 mqtt_test.h 和程序代码文件 mqtt_test.c,以及 entry.c 文件。完整代码如下。MQTT 服务器地址为 192.168.1.197。订阅主题为 WHCCNUSta1,发送的测试消息为"Temp:12,Humi:24"。

```
/* mqtt_test.h */
#ifndef __MQTT_TEST_H__
#define __MQTT_TEST_H__
void mqtt_test(void);
#endif

/* mqtt_test.c */

#include <stdio.h>
```

```c
#include <unistd.h>

#include "ohos_init.h"
#include "cmsis_os2.h"

#include <unistd.h>
#include "hi_wifi_api.h"
//#include "wifi_sta.h"
#include "lwip/ip_addr.h"
#include "lwip/netifapi.h"

#include "lwip/sockets.h"

#include "MQTTPacket.h"
#include "transport.h"

#define    MQTTserverIPADDR   "192.168.1.197"
#define    MQTTQoS            0
#define    MQTTID             "Monitor1"
#define    MQTTKeepAliveInt   20

int mqtt_connect(void)
{
        int toStop = 3;
        MQTTPacket_connectData data = MQTTPacket_connectData_initializer;
        int rc = 0;
        int mysock = 0;
        unsigned char buf[200];
        int buflen = sizeof(buf);
        int msgid = 1;
        MQTTString topicString = MQTTString_initializer;
        int req_qos = MQTTQoS;
        char * payload = "Temp: 12,Humi: 24";
        int payloadlen = strlen(payload);
        int len = 0;
        //char * host = "106.13.62.194";
        char * host = MQTTserverIPADDR;
        int port = 1883;


        mysock = transport_open(host, port);
        if(mysock < 0)
        return mysock;

        printf("Sending to hostname %s port %d\n", host, port);

        data.clientID.cstring = MQTTID;
        data.keepAliveInterval = MQTTKeepAliveInt;
        data.cleansession = 1;
        data.username.cstring = "";
```

```c
            data.password.cstring = "";

            len = MQTTSerialize_connect(buf, buflen, &data);
            rc = transport_sendPacketBuffer(mysock, buf, len);

            /* 等待 connack 包 */
            if (MQTTPacket_read(buf, buflen, transport_getdata) == CONNACK)
            {
                    unsigned char sessionPresent, connack_rc;

                    if (MQTTDeserialize_connack(&sessionPresent, &connack_rc, buf, buflen) != 1 || connack_rc != 0)
                    {
                            printf("Unable to connect, return code %d\n", connack_rc);
                            goto exit;
                    }
            }
            else
                goto exit;

            /* 注册 */
            topicString.cstring = "WHCCNUSta1";
            len = MQTTSerialize_subscribe(buf, buflen, 0, msgid, 1, &topicString, &req_qos);
            rc = transport_sendPacketBuffer(mysock, buf, len);
            if (MQTTPacket_read(buf, buflen, transport_getdata) == SUBACK)   /* 等待 suback 包 */
            {
                    unsigned short submsgid;
                    int subcount;
                    int granted_qos;

                    rc = MQTTDeserialize_suback(&submsgid, 1, &subcount, &granted_qos, buf, buflen);
                    if (granted_qos != 0)
                    {
                            printf("granted qos != 0, %d\n", granted_qos);
                            goto exit;
                    }
            }
            else
                goto exit;

            /* 循环获取注册主题的消息 msgs */
            //topicString.cstring = "pubtopic";
            topicString.cstring = "WHCCNUSta1";
            while (toStop)
            {
                    toStop = toStop - 1;
                    /* transport_getdata() 有内置的 1 秒超时 */
                    if (MQTTPacket_read(buf, buflen, transport_getdata) == PUBLISH)
                    {
                            unsigned char dup;
```

```c
                        int qos;
                        unsigned char retained;
                        unsigned short msgid;
                        int payloadlen_in;
                        unsigned char* payload_in;
                        int rc;
                        MQTTString receivedTopic;
                        rc = MQTTDeserialize_publish(&dup, &qos, &retained, &msgid, &receivedTopic,
                        &payload_in, &payloadlen_in, buf, buflen);
                        printf("message arrived %.*s\n", payloadlen_in, payload_in);

                        rc = rc;
                }

                printf("publishing reading\n");
                len = MQTTSerialize_publish(buf, buflen, 0, 0, 0, 0, topicString, (unsigned char*)payload, payloadlen);
                rc = transport_sendPacketBuffer(mysock, buf, len);
        }

        printf("disconnecting\n");
        len = MQTTSerialize_disconnect(buf, buflen);
        rc = transport_sendPacketBuffer(mysock, buf, len);
exit:
        transport_close(mysock);

        rc = rc;

        return 0;
}

void mqtt_test(void)
{
        mqtt_connect();
}
```

entry.c 文件用于创建线程。

```c
#include <stdio.h>

#include <unistd.h>
#include "ohos_init.h"
#include "cmsis_os2.h"
#include "wifi_con.h"
#include "mqtt_test.h"

static void* mqtt_task(void* argv)
{
        (void)argv;
        printf("\n MqttTask is running! Key = %d\n", value);
```

```c
/*该函数功能是连接WiFi,具体实现参考上一个实验。只有连接WiFi之后才能应用MQTT*/
NetCon();

    while ( true ) {
        mqtt_test();
        usleep(2000000);
    }
}

static void mqtt_taskentry(void)
{
    osThreadAttr_t attr;

    attr.name = "mqtttask";
    attr.attr_bits = 0U;
    attr.cb_mem = NULL;
    attr.cb_size = 0U;
    attr.stack_mem = NULL;
    attr.stack_size = 4096;
    attr.priority = 36;

    if (osThreadNew((osThreadFunc_t)mqtt_task, NULL, &attr) == NULL) {
        printf("[MQTTExample] Falied to create mqtttask!\n");
    }
}

SYS_RUN(mqtt_taskentry);
```

同一目录下的 BUILD.gn 文件内容如下。

```
static_library("mqtt_test") {
    sources = [
    "mqtt_test.c",
    "entry.c",
#下面两个文件用于连接WiFi
    "wifi_con.c",
    "wifi_connecter.c"
    ]

    include_dirs = [
    "//utils/native/lite/include",
    "//kernel/liteos_m/components/cmsis/2.0",
    "//base/iot_hardware/interfaces/kits/wifiiot_lite",
    "//base/iot_hardware/peripheral/interfaces/kits",
#V1.0 版本用该目录
#"//vendor/hisi/hi3861/hi3861/third_party/lwip_sack/include",
    "//device/hisilicon/hispark_pegasus/sdk_liteos/third_party/lwip_sack/include",
    #"//foundation/communication/interfaces/kits/wifi_lite/wifiservice",
    "//foundation/communication/wifi_lite/interfaces/wifiservice",
    "//third_party/pahomqtt/MQTTPacket/src",
    "//third_party/pahomqtt/MQTTPacket/samples",
```

```
#"//vendor\hisi\hi3861\hi3861\components\at\src",
"//device/hisilicon/hispark_pegasus/sdk_liteos/components/at/src"
        ]
}
```

上级目录下的 BUILD.gn 文件内容如下。

```
import("//build/lite/config/component/lite_component.gni")

lite_component("app") {
        features = [
        "mqtt_test: mqtt_test"
        ]
}
```

在 https://www.emqx.com/zh/downloads/broker/4.3.10/emqx-windows-4.3.10.zip 下载开源 MQTT 消息服务器 emqx，解压后，命令行下进入 bin 目录，通过"emqx start"启动服务器。在 https://packages.emqx.io/MQTTX/v1.6.2/MQTTX.Setup.1.6.2.exe 下载 MQTTX 客户端。

3. 实验结果

编译后运行，在串口可以看到如下输出。

```
ConnectToHotspot!
connect status: 0
Sending to hostname 192.168.1.197 port 1883
message arrived 123457
publishing reading
message arrived Temp: 12,Humi: 24
publishing reading
message arrived Temp: 12,Humi: 24
publishing reading
disconnecting
```

在 MQTTX 客户端可以收到从 Himself861 发送的订阅消息，如图 3.4 所示。

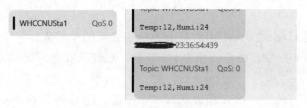

图 3.4　MQTTX 客户端

4. 扩展

MQTT 是机器对机器（Machine to Machine，M2M）/物联网（IoT）连接协议，是一个极其轻量级的发布/订阅消息传输协议。对于需要较小代码占用空间和/或网络带宽非常宝贵的远程连接非常有用，是专为受限设备和低带宽、高延迟或不可靠的网络而设计的协议。该协议成为 M2M 或 IoT 的理想选择。例如，通过卫星链路与代理通信的传感器、与医疗服务提供者的拨号连接，以及一系列家庭自动化和小型设备场景中使用了该协议。移动应用中，因为 MQTT 体积小，功耗低，数据包最小，并且可以有效地将信息分配给一个或多个接收器，也有广泛应用。

MQTT 协议具有以下主要的几项特性。

(1) 使用发布/订阅消息模式，提供一对多的消息发布，解除应用程序耦合。

(2) 对负载内容屏蔽的消息传输。

(3) 使用 TCP/IP 提供网络连接。

(4) 三种消息发布服务质量。

"至多一次"，消息发布完全依赖底层 TCP/IP 网络。会发生消息丢失或重复。这一级别可用于如下情况：环境传感器数据，丢失一次读记录无所谓，因为不久后还会有第二次发送。这一种方式主要适用于普通 APP 的推送，倘若智能设备在消息推送时未联网，推送过去没收到，再次联网也就收不到了。

"至少一次"，确保消息到达，但消息重复可能会发生。

"只有一次"，确保消息到达一次。在一些要求比较严格的计费系统中，可以使用此级别。在计费系统中，消息重复或丢失会导致不正确的结果。这种最高质量的消息发布服务还可以用于即时通信类的 APP 的推送，确保用户收到且只会收到一次。

(5) 小型传输，开销很小(固定长度的头部是 2B)，协议交换最小化，以降低网络流量。

(6) 使用 Last Will 和 Testament 特性通知有关各方客户端异常中断的机制。Last Will 即遗言，用于通知同一主题下的其他设备发送遗言的设备已经断开了连接。Testament 遗嘱机制，功能类似于 Last Will。

MQTT 中的术语如下。

(1) **网络连接**：指由底层传输协议提供给 MQTT 使用的架构。网络连接的底层传输协议能够连通客户端和服务端并能提供有序的、可靠的、双向字节流。

(2) **应用消息**：指通过 MQTT 在网络中传输的应用程序数据。当应用消息通过 MQTT 传输的时候会附加上质量服务(QoS)和话题名称。

(3) **客户端**：指使用 MQTT 的程序或设备。客户端总是去连接服务端。它可以发布其他客户端可能会感兴趣的应用消息、订阅自己感兴趣的应用消息、退订应用消息以及从服务端断开连接。

(4) **服务端**：是订阅或发布应用消息的客户端之间的中间人。一个服务端接受客户端的网络连接、接受客户端发布的应用消息、处理客户端订阅和退订的请求、转发匹配客户端订阅的应用消息。

(5) **订阅**：一个订阅由一个话题过滤器和一个最大的 QoS 组成。一个订阅只能关联一个会话。一个会话可以包含多个订阅。每个订阅都有不同的话题过滤器。

(6) **话题名称**：指附着于应用消息的标签，服务端用它来匹配订阅。服务端给每个匹配到的客户端发送一份应用信息的复制。

(7) **话题过滤器**：包含在订阅里的一个表达式，来表示一个或多个感兴趣的话题。话题过滤器可以包含通配符。

(8) **会话**：是一个有状态的客户端和服务端的交互。有些会话的存续依赖于网络连接，而其他则可以跨越一个客户端和服务端之间的多个连续的网络连接。

（9）**MQTT 控制包**：通过网络连接发送的包含一定信息的数据包。MQTT 规范定义了 14 个不同类型的控制包，其中一个（PUBLISH 包）用来传输应用信息。

实现 MQTT 协议需要客户端和服务器端通信完成。在通信过程中，MQTT 协议中有三种身份：发布者（Publish）、代理（Broker）（服务器）、订阅者（Subscribe）。其中，消息的发布者和订阅者都是客户端，消息代理是服务器，消息发布者可以同时是订阅者。

MQTT 传输的消息分为主题（Topic）和负载（Payload）两部分。Topic，可以理解为消息的类型，订阅者订阅（Subscribe）后，就会收到该主题的消息内容（Payload）。Payload 可以理解为消息的内容，是指订阅者具体要使用的内容。

MQTT 控制包结构包含三部分：固定包头，存在于所有 MQTT 控制包；可变包头，存在于某些 MQTT 控制包；载荷，存在于某些 MQTT 控制包。

第4章 应用UI开发实验

人生的目的,在发展自己的生命,可是也有为发展生命必须牺牲生命的时候。

因为平凡的发展,有时不如壮烈的牺牲足以延长生命的音响和光华。

绝美的风景,多在奇险的山川。

绝壮的音乐,多是悲凉的韵调。

高尚的生活,常在壮烈的牺牲中。

——李大钊

4.1 应用 UI 开发实验概览

1. 实验目的

本章实验的目的是掌握使用 HarmonyOS 通过类 Web 开发 UI 范式和声明式开发 UI 范式开发 UI 的基本方法,包括 UI 中的常用组件、页面路由、多组件 UI 开发、WebSocket 客户端和 MQTT 客户端等。

2. 实验设备

(1) 硬件设备为一台主流配置的计算机。

(2) 软件环境主要是 Windows 10 操作系统、DevEco Studio 3.0、HarmonyOS SDK 以及远程终端等。详细的环境配置和安装过程请参考第 1 章。

3. 实验内容

包含 22 个实验,分别为:

(1) 类 Web 开发 UI 组件 Input 实验。

(2) 类 Web 开发 UI 组件 Button 实验。

(3) 类 Web 开发 UI 组件 Form 实验。

(4) 类 Web 开发 UI 组件 Image 实验。

(5) 类 Web 开发 UI 组件 Picker 实验。

(6) 类 Web 开发 UI 组件 Tabs 实验。

(7) 页面路由实验。

(8) js2java-codegen 工具应用实验。
(9) 类 Web 开发 UI 实验。
(10) 声明式开发 UI 组件 Button 实验。
(11) 声明式开发 UI 组件 Text 实验。
(12) 声明式开发 UI 组件 Image 实验。
(13) 声明式开发 UI 组件 Slider 实验。
(14) 声明式开发 UI 组件 Flex 实验。
(15) 声明式开发 UI 组件 Stack 实验。
(16) 声明式开发 UI 组件 Tabs 实验。
(17) 声明式开发 UI 组件 List 实验。
(18) 声明式开发 UI 组件 Grid 实验。
(19) 声明式开发 UI 自定义组件实验。
(20) 声明式开发多组件 UI 实验。
(21) WebSocket 客户端实验。
(22) MQTT 客户端实验。

4. 实验过程

本章所有实验遵循以下步骤。
(1) 环境准备。
(2) 新建 APP 工程。
(3) 代码编写及资源准备。
(4) 在 Previewer 中预览 UI。
(5) 远程模拟器或本地模拟器中运行。
(6) 查看功能并测试功能。

4.2 类 Web 开发 UI 组件 Input 实验

1. 实验内容

本实验在不同的场景使用不同类型的 Input 输入框,完成信息录入。

2. 实验代码

在 DevEco Studio 3.0 中新建一个项目,在 Choose Your Ability Template 页选择 Empty Template,单击 Next 按钮进入 Config Your Project 页,在 Project name 框里填入项目名称,Project type 选择 Application,Development mode 选择 Traditional coding,Language 选择 JS。单击 Finish 按钮。将以下文件 xxx.hml 和 xxx.css 中所列代码分别替换掉该工程中文件 index.html 和 index.css 的内容。

代码如下。

```
<!-- xxx.hml -->
<div class = "container">
<div class = "label-item">
```

```html
<label>memorandum</label>
</div>
<div class="label-item">
<label class="lab" target="input1">内容：</label>
<input class="flex" id="input1" placeholder="Enter content" />
</div>
<div class="label-item">
<label class="lab" target="input3">日期：</label>
<input class="flex" id="input3" type="date" placeholder="Enter date" />
</div>
<div class="label-item">
<label class="lab" target="input4">时间：</label>
<input class="flex" id="input4" type="time" placeholder="Enter time" />
</div>
<div class="label-item">
<label class="lab" target="checkbox1">已完成：</label>
<input class="flex" type="checkbox" id="checkbox1" style="width: 100px; height: 100px; " />
</div>
<div class="label-item">
<input class="flex" type="button" id="button" value="保存" onclick="btnclick"/>
</div>
</div>
```

```css
/* xxx.css */
.container {
    flex-direction: column;
    background-color: #F1F3F5;
}
.label-item {
    align-items: center;
    border-bottom-width: 1px; border-color: #dddddd;
}
.lab {
    width: 400px; }
label {
    padding: 30px;
    font-size: 30px;
    width: 320px;
    font-family: serif;
    color: #9370d8;
    font-weight: bold;
}
.div-button {
    flex-direction: column;
    width: 100%;
    font-size: 30px;
}
.textareaPadding {
    padding-left: 100px;
}
```

3. 实验结果

在 DevEco Studio 3.0 中使用 Previewer 预览,得到如图 4.1 所示的 App 界面。

图 4.1 组件 Input 示例

4.3 类 Web 开发 UI 组件 Button 实验

1. 实验内容

本实验练习 Button 组件的各种类型。

2. 实验代码

在 DevEco Studio 3.0 中新建一个项目,在 Choose Your Ability Template 页选择 Empty Template,单击 Next 按钮进入 Config Your Project 页,在 Project name 框里填入项目名称,Project type 选择 Application,Development mode 选择 Traditional coding,Language 选择 JS。单击 Finish 按钮。将以下文件 xxx.hml、xxx.css 和 xxx.js 中所列代码分别替换掉该工程中文件 index.html、index.css 和 index.js 的内容。在 images 目录下建一个图片资源文件 paobu.png。

代码如下。

```
<!-- xxx.hml -->
<div class = "div-button">
<text class = "title">普通按钮</text>
<div class = "circleall">
<button class = "buttons">确认按钮</button>
<button class = "buttons" waiting = "true">Loading</button>
</div>
```

```html
<div class="circleall">
<button class="buttons icon" icon="/common/images/paobu.png" value="跑步" placement="start"></button>
<button class="buttons white">禁用按钮</button>
</div>
<div class="circleall">
<button class="buttons oriage">重置</button>
<button class="buttons warn">告警按钮</button>
</div>
<text class="title">胶囊形按钮</text>
<div class="circleall">
<button class="buttons" type="capsule">确认按钮</button>
<button class="buttons" type="capsule" waiting="true">Loading</button>
</div>
<div class="circleall">
<button class="buttons" type="capsule">清除按钮</button>
<button class="buttons white" type="capsule">禁用按钮</button>
</div>
<div class="circleall">
<button class="buttons oriage" type="capsule">重置按钮</button>
<button class="buttons warn" type="capsule">告警按钮</button>
</div>
<text class="title">圆形按钮</text>
<div class="circlealls">
<button class="circle" type="circle" icon="/common/images/paobu.png">icon按钮</button>
<button class="circle cir" type="circle" icon="/common/images/paobu.png">icon按钮</button>
</div>
<text class="title">文本按钮</text>
<div class="circleall">
<button class="text" type="text">文本按钮1</button>
<button class="text text1" type="text">文本按钮2</button>
<button class="text text2" type="text">文本按钮3</button>
</div>
<text class="title">长胶囊形按钮</text>
<div class="all">
<button class="download" type="download">下载按钮按钮</button>
<button class="download white" type="download">禁止下载按钮</button>
<button class="download" type="download" id="download-btn" onclick="setProgress">{{downloadText}}</button>
</div>
</div>
```

```css
/* xxx.css */
.div-button {
    flex-direction: column;
    width: 100%;
}
.circleall{
    width: 90%;
```

```css
        flex-direction: row;
        justify-content: space-around;
        margin-left: 4%;
}
.circlealls{
        flex-direction: row;
}
.buttons {
        margin-top: 15px;
        width: 45%;
        height: 45px;
        text-align: center;
        font-size: 14px;
        border-radius: 10px;
        background-color: #317aff;
}
.title{
        font-size: 13px;
        margin-top: 60px;
        margin-left: 20px;
        color: grey;
}
.all{
        flex-direction: column;
        align-items: center;
        margin-bottom: 20px;
}
.oriage{
        background-color: #ee8443;
}
.white{
        opacity: 0.4;
}
.icon{
        icon-height: 30px;
        icon-width: 30px;
}
.warn{
        background-color: #f55a42;
}
.circle {
        radius: 30px;
        icon-width: 30px;
        icon-height: 30px;
        margin-left: 20px;
        margin-top: 20px;
        background-color: #317aff;
}
.cir{
        background-color: #f55a42;
}
```

```css
.text {
    text-color: #0a59f7;
    font-size: 17px;
    font-weight: 600;
    font-family: sans-serif;
    font-style: normal;
}
.text1{
    text-color: #969696;
}
.text2{
    text-color: #e84026;
}
.download {
    margin-top: 15px;
    width: 88%;
    height: 45px;
    border-radius: 50px;
    text-color: white;
    background-color: #007dff;
}
```

```javascript
//xxx.js
export default {
    data: {
        progress: 10,
        downloadText: "进度条按钮"
    },
    setProgress(e) {
        var i = 0
        var set = setInterval(() =>{
            i += 10
            this.progress = i
            this.downloadText = this.progress + "%";
            this.$element('download-btn').setProgress({ progress: this.progress });
            if(this.progress >= 100){
                clearInterval(set)
                this.downloadText = "完成"
            }
        },1000)
    },
}
```

3. 实验结果

在 DevEco Studio 3.0 中使用 Previewer 预览,得到如图 4.2 所示的 App 界面。

图 4.2 组件 Button 示例

4.4 类 Web 开发 UI 组件 Form 实验

1. 实验内容

本实验练习 Form 组件,选择相应选项并提交或重置数据。

2. 实验代码

在 DevEco Studio 3.0 中新建一个项目,在 Choose Your Ability Template 页选择 Empty Template,单击 Next 按钮进入 Config Your Project 页,在 Project name 框里填入项目名称,Project type 选择 Application,Development mode 选择 Traditional coding,Language 选择 JS。单击 Finish 按钮。将以下文件 xxx.hml、xxx.css 和 xxx.js 中所列代码分别替换掉该工程中文件 index.html、index.css 和 index.js 的内容。

代码如下:

```
<!-- xxx.hml -->
<div class = "container">
<form onsubmit = "formSubmit" onreset = "formReset">
<text style = "font-size: 30px; margin-bottom: 20px; margin-top: 100px;">
<span> Form </span>
</text>
<div style = "flex-direction: column; width: 90%; padding: 30px 0px;">
<text class = "txt">选择至少一个选项</text>
<div style = "width: 90%; height: 150px; align-items: center; justify-content: space-around;">
```

```html
<label target="checkbox1">复选项1</label>
<input id="checkbox1" type="checkbox" name="checkbox1"></input>
<label target="checkbox2">复选项2</label>
<input id="checkbox2" type="checkbox" name="checkbox2"></input>
</div>
<divider style="margin: 20px 0px; color: pink; height: 5px;"></divider>
<text class="txt">选择一个选项</text>
<div style="width: 90%; height: 150px; align-items: center; justify-content: space-around;">
<label target="radio1">单选项1</label>
<input id="radio1" type="radio" name="myradio"></input>
<label target="radio2">单选项2</label>
<input id="radio2" type="radio" name="myradio"></input>
</div>
<divider style="margin: 20px 0px; color: pink; height: 5px;"></divider>
<text class="txt">文本框</text>
<input type="text" placeholder="输入内容" style="margin-top: 50px;"></input>
<div style="width: 90%; align-items: center; justify-content: space-between; margin: 40px;">
<input type="submit">提交</input>
<input type="reset">重置</input>
</div>
</div>
</form>
</div>
```

```css
/* index.css */
.container {
    flex-direction: column;
    align-items: center;
    background-color: #F1F3F5;
}
.txt {
    font-size: 33px;
    font-weight: bold;
    color: darkgray;
}
label{
    font-size: 20px;
}
```

```js
/* xxx.js */
import prompt from '@system.prompt';
export default {
    formSubmit() {
        prompt.showToast({
            message: 'Submited.'
        })
    },
    formReset() {
        prompt.showToast({
            message: 'Reset.'
        })
    }
}
```

3. 实验结果

在 DevEco Studio 3.0 中使用 Previewer 预览,得到如图 4.3 所示的 App 界面。

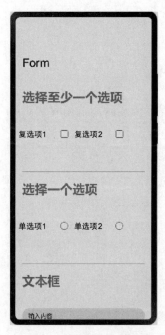

图 4.3 组件 Form 示例

4.5 类 Web 开发 UI 组件 Image 实验

1. 实验内容

在本实验中,练习使用组件 Image。

2. 实验代码

在 DevEco Studio 3.0 中新建一个项目,在 Choose Your Ability Template 页选择 Empty Template,单击 Next 按钮进入 Config Your Project 页,在 Project name 框里填入项目名称,Project type 选择 Application,Development mode 选择 Traditional coding,Language 选择 JS。单击 Finish 按钮。将以下文件 xxx.hml、xxx.css 和 xxx.js 中所列代码分别替换掉该工程中文件 index.html、index.css 和 index.js 的内容。

各个文件的代码如下。

```
<!-- xxx.hml -->
<div class = "page-container">
  <div class = "content">
    <div class = "image-container">
      <image class = "testimage" src = "{{testuri}}" style = "display: {{displaytype}}; opacity: {{imageopacity}};" onclick = "changedisplaytype" onlongpress = "changeopacity"></image>
    </div>
    <div class = "text-container">
```

```html
<text style = "font-size: 37px; font-weight: bold; color: orange; text-align: center;
width: 100%;">Touch and hold the image</text>
</div>
</div>
</div>
```

```css
/* xxx.css */
.page-container {
        flex-direction: column;
        align-self: center;
        justify-content: center;
        background-color: #F1F3F5;
        background-color: #F1F3F5;
}
.content{
        flex-direction: column;
}
.image-container {
        width: 100%;
        height: 300px;
        align-items: center;
        justify-content: center;
}
.text-container {
        margin-top: 50px;
        width: 100%;
        height: 60px;
        flex-direction: row;
        justify-content: space-between;
}
.testimage {
        width: 100%;
        height: 400px;
        object-fit: scale-down;
        border-radius: 20px;
}
```

```js
/* xxx.js */
import prompt from '@system.prompt';
export default {
        data: {
                testuri: 'common/images/bg-tv.jpg',
                imageopacity: 1,
                timer: null
        },
        changeopacity: function () {
                prompt.showToast({
                        message: 'Touch and hold the image.'
                })
                var opval = this.imageopacity * 20
                clearInterval(this.timer);
                this.timer = setInterval(()=>{
                        opval--;
```

```
                this.imageopacity = opval / 20
                if (opval === 0) {
                        clearInterval(this.timer)
                        this.imageopacity = 1
                }
        },100);
    }
}
```

3. 实验结果

在 DevEco Studio 3.0 中使用 Previewer 预览，得到如图 4.4 所示的 App 界面。

图 4.4　组件 Image 示例

长按图片后将慢慢隐藏图片，当完全隐藏后再重新显示原始图片。代码中的定时器 setInterval 每隔一段时间改变图片透明度，实现慢慢隐藏的效果，当透明度为 0 时清除定时器，设置透明度为 1。

4.6　类 Web 开发 UI 组件 Picker 实验

1. 实验内容

在本实验中，练习 Picker 组件的应用。该示例中，自定义填写当前的健康情况来进行打卡。

2. 实验代码

在 DevEco Studio 3.0 中新建一个项目，在 Choose Your Ability Template 页选择 Empty Template，单击 Next 按钮进入 Config Your Project 页，在 Project name 框里填入项目名称，Project type 选择 Application，Development mode 选择 Traditional coding，Language 选择 JS。单击 Finish 按钮。将以下文件 xxx.hml、xxx.css 和 xxx.js 中所列代

码分别替换掉该工程中文件 index.html、index.css 和 index.js 的内容。

代码如下。

```html
<!-- xxx.hml -->
<div class="doc-page">
    <text class="title">Health check-in</text>
    <div class="out-container">
        <text class="txt">办公地点：</text>
        <picker class="pick" focusable="true" type="text" value="{{pos}}" range="{{posarr}}" onchange="setPos"></picker>
    </div>
    <divider class="dvd"></divider>
    <div class="out-container">
        <text class="txt">工作时间：</text>
        <picker class="pick" type="date" value="{{datevalue}}" start="2002-2-5" end="2030-6-5" selected="{{dateselect}}" lunarswitch="true" onchange="dateonchange"></picker>
    </div>
    <divider class="dvd"></divider>
    <div class="out-container">
        <text class="txt">是否有发烧感冒症状？</text>
        <picker class="pick" type="text" value="{{yorn1}}" range="{{yesno}}" selected="1" onchange="isFever"></picker>
    </div>
    <divider class="dvd"></divider>
    <div class="out-container">
        <text class="txt">和COVID确诊患者有没有近距离接触？</text>
        <picker class="pick" type="text" value="{{yorn2}}" range="{{yesno}}" selected="1" onchange="isTouch"></picker>
    </div>
    <div class="out-container">
        <button value="提交" style="margin-top: 100px; width: 50%; font-color: #0000ff; height: 80px" onclick="showtoast"></button>
    </div>
</div>
```

```css
/* xxx.css */
.doc-page {
    flex-direction: column;
    background-color: #F1F3F5;
}
.title {
    margin-top: 30px;
    margin-bottom: 30px;
    margin-left: 50px;
    font-weight: bold;
    color: #0000ff;
    font-size: 38px;
}
```

```css
.out-container {
        flex-direction: column;
        align-items: center;
}
.pick {
        width: 80%;
        height: 40px;
        border: 1px solid #0000ff;
        border-radius: 20px;
        padding-left: 12px;
}
.txt {
        width: 80%;
        font-size: 18px;
        text-align: left;
        margin-bottom: 12px;
        margin-left: 12px;
}
.dvd {
        margin-top: 30px;
        margin-bottom: 30px;
        margin-left: 80px;
        margin-right: 80px;
        color: #6495ED;
        stroke-width: 6px;
}
```

```js
//xxx.js
import pmt from '@system.prompt'
export default {
        data: {
                yorn1: '否',
                yorn2: '否',
                pos: '居家',
                yesno: ['是', '否'],
                posarr: ['居家', '公司'],
                datevalue: '选择时间',
                datetimeselect: '2022-5-6-11-25',
                dateselect: '2022-9-17',
                showbuild: true
        },
        onInit() {
        },
        isFever(e) {
                this.yorn1 = e.newValue
        },
        isTouch(e) {
                this.yorn2 = e.newValue
        },
        setPos(e) {
```

```
                this.pos = e.newValue
                if (e.newValue === 'Non-research center') {
                    this.showbuild = false
                } else {
                    this.showbuild = true
                }
            },
            setbuild(e) {
                this.build = e.newValue
            },
            dateonchange(e) {
                e.month = e.month + 1;
                this.datevalue = e.year + "-" + e.month + "-" + e.day;
                pmt.showToast({ message: "date: " + e.year + "-" + e.month + "-" + e.day })
            },
            showtoast() {
                pmt.showToast({
                    message: '已提交.',
                    duration: 2000,
                    gravity: 'center'
                })
            }
        }
```

3. 实验结果

在 DevEco Studio 3.0 中使用 Previewer 预览，得到如图 4.5 所示的 App 界面。

图 4.5　组件 Picker 示例

滑动选择器组件 Picker 类型支持普通选择器、日期选择器、时间选择器、时间日期选择器和多列文本选择器。该实验中,使用了日期选择器和普通选择器。普通选择器设置取值范围时,需要使用数据绑定的方式。如"办公地点"选择器,绑定数据为 posarr。该变量在 index.js 中定义。

4.7 类 Web 开发 UI 组件 Tabs 实验

1. 实验内容

在本实验中,练习 Tabs 组件的使用。

2. 实验代码

在 DevEco Studio 3.0 中新建一个项目,在 Choose Your Ability Template 页选择 Empty Template,单击 Next 按钮进入 Config Your Project 页,在 Project name 框里填入项目名称,Project type 选择 Application,Development mode 选择 Traditional coding,Language 选择 JS。单击 Finish 按钮。将以下文件 xxx.hml、xxx.css 和 xxx.js 中所列代码分别替换掉该工程中文件 index.html、index.css 和 index.js 的内容。在 images 目录下建两个图片资源文件 lenacolor.png 和 img2.png。

完整代码如下。

```
<!-- xxx.hml -->
<div class = "container">
<tabs onchange = "changeTabactive">
<tab-content>
<div class = "item-container" for = "datas.list">
<div if = "{{ $item.title == 'List1'?true: false}}">
<image src = "common/images/bg-tv.jpg" style = "object-fit: contain;"></image>
</div>
<div if = "{{ $item.title == 'List2'?true: false}}">
<image src = "common/images/lenacolor.png" style = "object-fit: none;"></image>
</div>
<div if = "{{ $item.title == 'List3'?true: false}}">
<image src = "common/images/img2.png" style = "object-fit: contain;"></image>
</div>
</div>
</tab-content>
<tab-bar class = "tab_bar mytabs" mode = "scrollable">
<div class = "tab_item" for = "datas.list">
<text style = "color: {{ $item.color}}; ">{{ $item.title}}</text>
<div class = "underline-show" if = "{{ $item.show}}"></div>
<div class = "underline-hide" if = "{{! $item.show}}"></div>
</div>
</tab-bar>
</tabs>
</div>
/* xxx.css */
.container{
```

```css
        background-color: #F1F3F5;
}
.tab_bar {
        width: 100%;
}
.tab_item {
        flex-direction: column;
        align-items: center;
}
.tab_item text {
        font-size: 32px;
}
.item-container {
        justify-content: center;
        flex-direction: column;
}
.underline-show {
        height: 2px;
        width: 160px;
        background-color: #FF4500;
        margin-top: 7.5px;
}
.underline-hide {
        height: 2px;
        margin-top: 7.5px;
        width: 160px;
}
```

```js
/* xxx.js */
import prompt from '@system.prompt';
export default {
        data() {
                return {
                        datas: {
                                color_normal: '#878787',
                                color_active: '#ff4500',
                                show: true,
                                list: [{
                                        i: 0,
                                        color: '#ff4500',
                                        show: true,
                                        title: 'List1'
                                }, {
                                        i: 1,
                                        color: '#878787',
                                        show: false,
                                        title: 'List2'
                                }, {
                                        i: 2,
                                        color: '#878787',
                                        show: false,
```

```
                            title: 'List3'
                        }]
                    }
                }
            },
            changeTabactive (e) {
                for (let i = 0; i < this.datas.list.length; i++) {
                    let element = this.datas.list[i];
                    element.show = false;
                    element.color = this.datas.color_normal;
                    if (i === e.index) {
                        element.show = true;
                        element.color = this.datas.color_active;
                    }
                }
            }
        }
```

3. 实验结果

在 DevEco Studio 3.0 中使用 Previewer 预览，得到如图 4.6 所示的 App 界面。

图 4.6　组件 Tabs 示例

单击标签切换内容，选中 List2 后标签文字颜色变红，并显示下画线和该页的图片。用 tabs、tab-bar 和 tab-content 实现单击切换功能，再定义数组，设置属性。使用 change 事件改变数组内的属性值实现变色及下画线的显示。

4.8　页面路由实验

1. 实验内容

页面路由 router 根据页面的 URI 找到目标页面，从而实现跳转。本实验实现两个页面之间的跳转。

2. 实验代码

在 DevEco Studio 3.0 中新建一个项目，在 Choose Your Ability Template 页选择 Empty Template，单击 Next 按钮进入 Config Your Project 页，在 Project name 框里填入项目名称，Project type 选择 Application，Development mode 选择 Traditional coding，Language 选择 JS。单击 Finish 按钮。将以下文件 index.html、index.css 和 index.js 中的代码替代新建工程中的文件。

```html
<!-- index.hml -->
<div class="container">
    <text class="title">这是索引页。</text>
    <button type="capsule" value="转到下页" class="button" onclick="launch"></button>
</div>
```

```css
/* index.css */
.container {
    flex-direction: column;
    justify-content: center;
    align-items: center;
}

.title {
    font-size: 50px;
    margin-bottom: 50px;
}
.button{
    font-size: 35px;
}
```

```js
// index.js
import router from '@system.router';
export default {
    launch() {
        router.push({
            uri: 'pages/detail/detail',
        });
    },
}
```

在 Project 窗口，打开 entry→src→main→js→default，右击 pages 目录，在弹出的快捷菜单中选择 New→JS Page 命令，创建一个 detail 页。index 和 detail 这两个页面均包含一个 text 组件和 button 组件。text 组件用来指明当前页面，button 组件用来实现两个页面之间的相互跳转。

```html
<!-- detail.hml -->
<div class="container">
    <text class="title">这是详情页。</text>
    <button type="capsule" value="返回" class="button" onclick="launch"></button>
</div>
```

index 和 detail 页面的页面样式，text 组件和 button 组件居中显示，两个组件之间间距为 50px。

```css
/* detail.css */
.container {
    flex-direction: column;
    justify-content: center;
    align-items: center;
}

.title {
    font-size: 50px;
    margin-bottom: 50px;
}
```

index 页面调用 router.push()路由到 detail 页。detail 页调用 router.back()回到 index 页。

```js
//detail.js
import router from '@system.router';
export default {
    launch() {
        router.back();
    },
}
```

3. 实验结果

在 DevEco Studio 3.0 中使用 Previewer 预览，得到如图 4.7 所示的 App 界面。

图 4.7　页面路由示例

页面的 JS 文件中实现跳转逻辑。调用 router.push()接口将 URI 指定的页面添加到路由栈中，即跳转到 URI 指定的页面。在调用 router 方法之前，需要导入 router 模块。

4.9 js2java-codegen 工具应用实验

使用基于 JavaScript 扩展的类 Web 开发范式的方舟开发框架提供了 JS FA 调用 Java PA 的机制,该机制提供了一种通道来传递方法调用、处理数据返回以及订阅事件上报。

当前提供 Ability 和 Internal Ability 两种调用方式,根据业务场景选择合适的调用方式进行开发。

Ability 调用方式下,Java PA 拥有独立的 Ability 生命周期,JS FA 使用远端进程通信拉起并请求 PA 服务,适用于基本服务供多 FA 调用或者服务在后台独立运行的场景。Internal Ability 调用方式下,Java PA 与 FA 共进程,采用内部函数调用的方式和 FA 进行通信,适用于对服务响应时延要求较高的场景。该方式下 PA 不支持其他 FA 访问调用。

对于 Internal Ability 调用方式的开发,可以使用 js2java-codegen 工具自动生成代码,提高开发效率。

1. 实验内容

本实验通过 js2java-codegen 工具生成模板代码。包含 Java 代码和 JavaScript 代码。

2. 实验过程

js2java-codegen 工具生成的模板代码包含 Java 代码和 JavaScript 代码。其中,Java 代码会被直接编译成字节码文件,并且对应 Ability 类中会被自动添加注册与反注册语句;JavaScript 代码需要手动调用,因此需要在编译前设置好 JavaScript 代码的输出路径。

js2java-codegen 工具通过注解来获取信息并生成所需的代码,有以下注解的三种用法。

(1) @InternalAbility 注解。

类注解,被用于 InternalAbility 的、包含实际业务代码的类(简称 InternalAbility 类)。只支持文件中 public 的顶层类,不支持接口类和注解类。包含一个参数 registerTo,值为需要注册到的 Ability 类全名。

例如,代表 Service 类是一个 InternalAbility 类,注册到位于 com.example 包中的、名为 Ability 的 Ability 类。

```
@InternalAbility(registerTo = "com.example.Ability")
public class Service{}
```

(2) @ExportIgnore 注解。

方法注解,用于 InternalAbility 类中的某些方法,表示该方法不暴露给 JavaScript 侧来调用。仅对 public 方法有效。

例如,以下代码示意 service() 方法不会被暴露给 JavaScript 侧。

```
@ExportIgnore
    public int service(int input) {
        return input;
    }
```

(3) @ContextInject 注解。

用于 AbilityContext 上的注解。该类由 HarmonyOS 的 Java API 提供,开发者可通过

它获取 API 中提供的信息。

例如，以下代码代表开发者可以借助 abilityContext 对象获取 API 中提供的信息。

```
@ContextInject
    AbilityContext abilityContext;
```

使用 DevEco Studio 新建一个包含 JavaScript 前端的简单手机项目，用于 FA 调用 PA。

在需要进行代码生成的模块下的 build.gradle 中控制开关和进行编译设置。想要快速验证功能，可选择修改 entry 模块的 build.gradle，通过 entry 模块进行验证。

编译参数位于 ohos -> defaultConfig 中，只需添加如下设置即可。需在此处设置 JavaScript 模板代码生成路径，即 jsOutputDir 对应的值。

```
//在文件头部定义 JavaScript 模板代码生成路径
def jsOutputDir = project.file("src/main/js/default/generated").toString()

//在 ohos -> defaultConfig 中设置 JavaScript 模板代码生成路径
javaCompileOptions {
    annotationProcessorOptions{
        arguments = ["jsOutputDir": jsOutputDir] //JavaScript 模板代码生成赋值
    }
}
```

工具开关位于 ohos 中，只需添加如下设置即可。值设为 true 则启用工具，值为 false 或不进行配置则不启用工具。

```
compileOptions {
    f2pautogenEnabled true              //此处为启用 js2java-codegen 工具的开关
}
```

模板代码的生成需要提供用于 FA 调用的 PA，因此需要自己编写 InternalAbility 类，然后在类上加 @InternalAbility 注解，registerTo 参数设为想要注册到的 Ability 类的全称（Ability 类可使用项目中已有的 MainAbility 类，或创建新的 Ability 类）。

注意 InternalAbility 类中需要暴露给 FA 来调用的方法只能是 public 类型的非静态非 void() 方法，如若不是则不会被暴露。

一个简单的 InternalAbility 类实现如下，文件名为 Service.java，与 MainAbility 类同包，用注解注册到 MainAbility 类。类里面包含一个 add() 方法作为暴露给 JavaScript FA 来调用的能力，实现了两数相加的功能，入参为两个 int 参数，返回值为两数的和。

```
package com.example.myapplication;

import ohos.annotation.f2pautogen.InternalAbility;

@InternalAbility(registerTo = "com.example.myapplication.MainAbility")
                        /* 此处 registerTo 的参数为项目中 MainAbility 类的全称 */
public class Service {
    public int add(int num1, int num2) {
        return num1 + num2;
    }
}
```

单击菜单栏中的 Build -> Build HAP(s)/APP(s) -> Build HAP(s)，即可完成对项目的编译，同时 js2java-codegen 工具会在编译过程中完成 FA 调用 PA 通道的建立。

编译过程会生成 Java 和 JavaScript 的模板代码。其中，JavaScript 的模板代码位于开发者在编译设置中设置的路径，名称与 InternalAbility 类的名称相对应；而 Java 的模板代码位于 entry→build→generated→source→annotation→debug→InternalAbility 类同名包→InternalAbility 类名＋Stub.java，而该类的调用语句会被注入到 MainAbility 类的字节码当中。

为了简易直观地检验工具生成的代码的可用性，可通过修改 entry→src→main→js→default→pages→index→index.js 来调用 Java 侧的能力并在前端页面展示效果。

可通过 import 方式引入 JavaScript 侧 FA 接口，例如 import Service from '../../generated/Service.js'; from 后的值需要与编译设置中的路径进行统一。生成的 JavaScript 代码文件名及类名与 InternalAbility 类名相同。

一个简单的 index.js 页面实现如下，调用了 JavaScript 侧接口，传入了 1 和 10 两个参数，并把返回的结果打印在 title 中，这样只要运行该应用就可以验证 FA 调用 PA 是否成功。

```javascript
import Service from '../../generated/Service.js';    /* 此处 FA 路径和类名对应之前的 jsOutput
                                                        路径以及 InternalAbility 的名字 */
export default {
    data: {
        title: "Result: "
    },
    onInit() {
        const echo = new Service();         //此处新建 FA 实例
        echo.add(1,10)
        .then((data) => {
            //this.title += data["abilityResult"];
                                            //此处取到运算结果，并加到 title 之后
        });
    }
}
```

同目录下的 index.html 只显示 title 的内容。

```html
<div class = "container">
<text class = "title">
{{ title }}
</text>
</div>
```

3. 实验结果

启动手机模拟器，启动成功后运行应用，如图 4.8 所示。

说明 js2java-codegen 工具生成了有效的模板代码，成功地建立起了 FA 调用 PA 的通道。

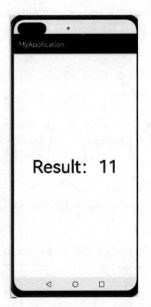

图 4.8　使用 js2java-codegen 工具生成模板示例

4.10　类 Web 开发 UI 实验

1. 实验内容

在本实验中,练习如何用类 Web 开发范式开发一个 JS FA 应用。

2. 实验代码

在 DevEco Studio 3.0 中新建一个项目,在 Choose Your Ability Template 页选择 Empty Template,单击 Next 按钮进入 Config Your Project 页,在 Project name 框里填入项目名称,Project type 选择 Application,Development mode 选择 Traditional coding,Language 选择 JavaScript。单击 Finish 按钮。

在 index.hml 文件中构建页面布局,一共分成三个部分,标题区、展示区和详情区。标题区较为简单,由两个按列排列的 text 组件构成。展示区由包含四个 image 组件的 swiper 组件构成。详情区由 image 组件和 text 组件构成。图片资源放置到 common 目录下。

文件 index.hml 代码如下。

```
<!-- index.hml -->
<div class = "container">
<!-- 标题区 -->
<div class = "title">
<text class = "name">书籍</text>
<text class = "sub-title">选择</text>
</div>
<div class = "display-style">
<!-- 展示区 -->
<swiper id = "swiperImage" class = "swiper-style">
```

```html
<image src = "{{ $item}}" class = "image-mode" focusable = "true" for = "{{imageList}}">
</image>
</swiper>
<!-- 产品详情区 -->
<div class = "container">
<div class = "selection-bar-container">
<div class = "selection-bar">
<image src = "{{ $item}}" class = "option-mode" onfocus = "swipeToIndex({{ $idx}})" onclick = "swipeToIndex({{ $idx}})" for = "{{imageList}}"></image>
</div>
</div>
<div class = "description-first-paragraph">
<text class = "description">{{descriptionFirstParagraph}}</text>
</div>
<div class = "cart">
<text class = "{{cartStyle}}" onclick = "addCart" onfocus = "getFocus" onblur = "lostFocus" focusable = "true">{{cartText}}</text>
</div>
</div>
</div>
</div>
```

index.css 文件中通过 media query 管控手机和 TV 不同页面样式。此外,该页面样式还采用了 CSS 伪类的写法,当单击时或者焦点移动到 image 组件上,image 组件由半透明变成不透明,以此来实现选中的效果。该文件代码如下。

```css
.container {
        flex-direction: column;
}
/* TV 页面 */
@media screen and (device-type: tv) {
        .title {
                align-items: flex-start;
                flex-direction: column;
                padding-left: 60px;
                padding-right: 160px;
                margin-top: 15px;
        }

        .name {
                font-size: 20px;
        }

        .sub-title {
                font-size: 15px;
                color: #7a787d;
                margin-top: 10px;
        }

        .swiper-style {
                height: 300px;
```

```css
        width: 350px;
        indicator-color: #4682b4;
        indicator-selected-color: #f0e68c;
        indicator-size: 10px;
        margin-left: 50px;
}

.image-mode {
        object-fit: contain;
}

.selection-bar {
        flex-direction: row;
        align-content: center;
        margin-top: 20px;
        margin-left: 10px;
}

.option-mode {
        height: 40px;
        width: 40px;
        margin-left: 50px;
        opacity: 0.5;
        border-radius: 20px;
}

.option-mode: focus {
        opacity: 1;
}

.description-first-paragraph {
        padding-left: 60px;
        padding-right: 60px;
        padding-top: 30px;
}

.description {
        color: #7a787d;
        font-size: 15px;
}

.cart {
        padding-left: 60px;
        margin-top: 30px;
}

.cart-text {
        font-size: 20px;
        text-align: center;
        width: 300px;
        height: 50px;
```

```css
            background-color: #6495ed;
            color: white;
        }

        .cart-text-focus {
            font-size: 20px;
            text-align: center;
            width: 300px;
            height: 50px;
            background-color: #4169e1;
            color: white;
        }

        .add-cart-text {
            font-size: 20px;
            text-align: center;
            width: 300px;
            height: 50px;
            background-color: #ffd700;
            color: white;
        }
}

/* 手机页面 */
@media screen and (device-type: phone) {
        .title {
            align-items: flex-start;
            flex-direction: column;
            padding-left: 60px;
            padding-right: 160px;
            padding-top: 20px;
        }

        .name {
            font-size: 50px;
            color: #000000;
        }

        .sub-title {
            font-size: 30px;
            color: #7a787d;
            margin-top: 10px;
        }

        .display-style {
            flex-direction: column;
            align-items: center;
        }

        .swiper-style {
            height: 600px;
```

```css
        indicator-color: #4682b4;
        indicator-selected-color: #ffffff;
        indicator-size: 20px;
        margin-top: 15px;
}

.image-mode {
        object-fit: contain;
}

.selection-bar-container {
        height: 90px;
        justify-content: center;
}

.selection-bar {
        height: 90px;
        width: 500px;
        margin-top: 30px;
        justify-content: center;
        align-items: center;
}

.option-mode {
        object-fit: contain;
        opacity: 0.5;
}

.option-mode:active {
        opacity: 1;
}

.description {
        color: #7a787d;
}

.description-first-paragraph {
        padding-left: 60px;
        padding-top: 50px;
        padding-right: 60px;
}

.color-column {
        flex-direction: row;
        align-content: center;
        margin-top: 20px;
}

.color-item {
        height: 50px;
        width: 50px;
```

```css
            margin-left: 50px;
            padding-left: 10px;
        }

        .cart {
            justify-content: center;
            margin-top: 30px;
        }

        .cart-text {
            font-size: 35px;
            text-align: center;
            width: 600px;
            height: 100px;
            background-color: #6495ed;
            color: white;
        }

        .add-cart-text {
            font-size: 35px;
            text-align: center;
            width: 600px;
            height: 100px;
            background-color: #ffd700;
            color: white;
        }
}
```

index.js 文件中构建页面逻辑,主要实现的是两个逻辑功能。一个是当单击时或者焦点移动到不同的缩略图,swiper 滑动到相应的图片;另一个是当焦点移动到购物车区时,Add To Cart 背景颜色从浅蓝变成深蓝,单击后文字变化为"Cart + 1",背景颜色由深蓝色变成黄色。添加购物车不可重复操作。逻辑页面代码示例如下。

```javascript
// index.js
export default {
    data: {
        cartText: '添加到购物车',
        cartStyle: 'cart-text',
        isCartEmpty: true,
        descriptionFirstParagraph: '欢迎购买',
        imageList: ['/common/book1.PNG', '/common/book2.PNG', '/common/book3.PNG', '/common/book4.PNG'],
    },

    swipeToIndex(index) {
        this.$element('swiperImage').swipeTo({index: index});
    },

    addCart() {
        if (this.isCartEmpty) {
```

```
                    this.cartText = 'Cart + 1';
                    this.cartStyle = 'add-cart-text';
                    this.isCartEmpty = false;
            }
        },
        getFocus() {
            if (this.isCartEmpty) {
                    this.cartStyle = 'cart-text-focus';
            }
        },
        lostFocus() {
            if (this.isCartEmpty) {
                    this.cartStyle = 'cart-text';
            }
        },
}
```

3. 实验结果

该应用运行时界面如图 4.9 所示。

图 4.9　图书购买 App

此应用相对于 Hello World 应用模板,有更复杂的页面布局、页面样式和页面逻辑。该应用通过 media query 同时适配了手机和 TV,通过单击或者将焦点移动到书籍的缩略图来选择不同的书籍图片。

4.11　声明式开发 UI 组件 Button 实验

1. 实验内容

本实验练习声明式开发 UI 组件 Button。

2. 实验代码

在 DevEco Studio 3.0 中新建一个项目,在 Choose Your Ability Template 页选择 Empty Template,单击 Next 按钮进入 Config Your Project 页,在 Project name 框里填入项目名称,Project type 选择 Application,Development mode 选择 Traditional coding,Language 选择 eTS。单击 Finish 按钮。

将以下代码替换掉该工程中 index.ets 文件中的内容。在目录 media 中创建三个图片文件,分别是 loading.png、ic_public_app_filled.png 和 media.ic_public_delete_filled.png。

代码如下。

```
@Entry
@Component
struct ButtonExample {
    build() {
        Flex({ direction: FlexDirection.Column, alignItems: ItemAlign.Start, justifyContent: FlexAlign.SpaceBetween }) {
            Text('普通按钮').fontSize(19).fontColor(0xCCCCCC)
            Flex({ alignItems: ItemAlign.Center, justifyContent: FlexAlign.SpaceBetween }) {
                Button('确认按钮', { type: ButtonType.Normal, stateEffect: true }).borderRadius(8).backgroundColor(0x317aff).width(90)
                Button({ type: ButtonType.Normal, stateEffect: true }) {
                    Row() {
                        Image($r('app.media.loading')).width(20).height(20).margin({ left: 12 })
                        Text('加载中').fontSize(12).fontColor(0xffffff).margin({ left: 5, right: 12 })
                    }.alignItems(VerticalAlign.Center)
                }.borderRadius(8).backgroundColor(0x317aff).width(90)
                Button('禁用', { type: ButtonType.Normal, stateEffect: false }).opacity(0.5)
                    .borderRadius(8).backgroundColor(0x317aff).width(90)
            }

            Text('胶囊形按钮').fontSize(19).fontColor(0xCCCCCC)
            Flex({ alignItems: ItemAlign.Center, justifyContent: FlexAlign.SpaceBetween }) {
                Button('确认按钮', { type: ButtonType.Capsule, stateEffect: true }).backgroundColor(0x317aff).width(90)
                Button({ type: ButtonType.Capsule, stateEffect: true }) {
                    Row() {
                        Image($r('app.media.loading')).width(20).height(20).margin({ left: 12 })
                        Text('加载中').fontSize(12).fontColor(0xffffff).margin({ left: 5, right: 12 })
                    }.alignItems(VerticalAlign.Center).width(90)
                }.backgroundColor(0x317aff)
                .onClick((event: ClickEvent) => {
                    AlertDialog.show({ message: '登录成功' })
                })
```

```
                    Button('禁用', { type: ButtonType.Capsule, stateEffect: false
}).opacity(0.5)
                        .backgroundColor(0x317aff).width(90)
                }
                Text('圆形按钮').fontSize(19).fontColor(0xCCCCCC)
                Flex({ alignItems: ItemAlign.Center, wrap: FlexWrap.Wrap }) {
                    Button({ type: ButtonType.Circle, stateEffect: true }) {
                        Image($r('app.media.ic_public_app_filled')).width
(20).height(20)
                    }.width(55).height(55).backgroundColor(0x317aff)
                    Button({ type: ButtonType.Circle, stateEffect: true }) {
                        Image($r('app.media.ic_public_delete_filled')).width
(30).height(30)
                    }.width(55).height(55).margin({ left: 20 }).backgroundColor
(0xF55A42)
                }
            }.height(400).padding({ left: 35, right: 35, top: 35 })
        }
    }
```

3. 实验结果

在 DevEco Studio 3.0 中使用 Previewer 预览，得到如图 4.10 所示的 App 界面。

图 4.10　eTS 组件 Button 示例

展示了普通按钮、胶囊形按钮和圆形按钮三类按钮，以及按钮的 onClick 事件。

4. 扩展

Button 按钮组件类型包括普通按钮、胶囊形按钮和圆形按钮，通常用于响应单击操作。eTS 中通过调用两种形式的接口来创建 Button 按钮。

（1）包含子组件的按钮。

接口为 Button(options?：type?：ButtonType，stateEffect?：boolean)，其中，type 用于设置 Button 类型，stateEffect 属性设置 Button 是否开启单击效果。示例代码如下：

```
Button({ type: ButtonType.Normal, stateEffect: true }) {
    Row() {
        Image($r('app.media.loading')).width(20).height(20).margin({ left: 12 })
```

```
                Text('加载中').fontSize(12).fontColor(0xffffff).margin({ left: 5, right:
12 })
        }.alignItems(VerticalAlign.Center)
}.borderRadius(8).backgroundColor(0x317aff).width(90)
```

这段代码创建 Normal 类型的按钮，包含两个子组件，分别是 Image 和 Text。

（2）不包含子组件的按钮。

接口为 Button（label?：string，options?：   type?：ButtonType, stateEffect?：boolean），其中，label 确定所创建的 Button 是否包含子组件，type 用于设置 Button 类型，stateEffect 属性设置 Button 是否开启单击效果。示例代码如下。

```
Button('确认', { type: ButtonType.Capsule, stateEffect: true }).backgroundColor(0x317aff).width(90)
```

这段代码创建 Capsule 类型的按钮，不包含子组件。

Button 有三种可选类型，分别为 Capsule（胶囊形按钮）、Circle（圆形按钮）和 Normal（普通按钮），通过 type 进行设置。

当用于交互操作时，可以绑定 onClick 事件响应单击操作后的动作。示例代码如下。

```
Button('提交', { type: ButtonType.Normal, stateEffect: true })
.onClick(() =>{
        console.info('正在提交!')
})
```

如下代码示例了如何在 List 容器里边通过单击按钮进行页面跳转。

```
import router from '@ohos.router'

@Entry
@Component
struct ButtonCase1 {
        build() {
                List({ space: 4 }) {
                        ListItem() {
                                Button("页面一").onClick(() => {
                                        router.push({ url: 'xxx' })
                                })
                        }

                        ListItem() {
                                Button("页面二").onClick(() => {
                                        router.push({ url: 'yyy' })
                                })
                        }

                        ListItem() {
                                Button("页面三").onClick(() => {
                                        router.push({ url: 'zzz' })
                                })
                        }
```

```
            }
                .listDirection(Axis.Vertical)
                .backgroundColor(0xDCDCDC).padding(20)
        }
}
```

下面的代码示例了在用户登录页面,用户登录的提交操作。

```
@Entry
@Component
struct ButtonCase2 {
    build() {
        Column() {
            TextInput({ placeholder: '输入用户名' }).margin({ top: 20 })
            TextInput({ placeholder: '输入密码'})
                .type(InputType.Password).margin({ top: 20 })
            Button('注册').width(300).margin({ top: 20 })
        }.padding(20)
    }
}
```

4.12 声明式开发 UI 组件 Text 实验

1. 实验内容

本实验练习声明式开发 UI 组件 Text,该组件用于呈现一段文本信息。

2. 实验代码

在 DevEco Studio 3.0 中新建一个项目,在 Choose Your Ability Template 页选择 Empty Template,单击 Next 按钮进入 Config Your Project 页,在 Project name 框里填入项目名称,Project type 选择 Application,Development mode 选择 Traditional coding,Language 选择 eTS。单击 Finish 按钮。将以下代码替换掉该工程中 index.ets 文件中的内容。

代码如下。

```
@Entry
@Component
struct TextExample {
    build() {
            Flex({ direction: FlexDirection.Column, alignItems: ItemAlign.Start,
justifyContent: FlexAlign.SpaceBetween }) {
                Text('行高示例').fontSize(19).fontColor(0xCCCCCC)
                Text('设置行高的文本,可以看到行高更高了。远看山有色,近听水无声。
春去花还在,人来鸟不惊。')
                    .lineHeight(25).fontSize(12).border({ width: 1 }).padding(10)

                Text('文本溢出示例').fontSize(19).fontColor(0xCCCCCC)
                Text('这是一首画作欣赏诗,从诗中的描述来看,画中有山、水、花、鸟都
```

是典型的中国画题材,而且肯定是一幅画得相当逼真、传神的作品。作者通过文字的描述,把一幅本是静止的画变成了一幅美丽的风景卷轴展现出来:苍翠的山,流动的水,绽放的花,欢鸣的鸟,一派鲜活的景象,把读者引入了无限的遐想之中。当读者从遐想中回到现实的时候,才发现,画中的一切不过是一个个无生命的静物。')
        .textOverflow({ overflow: TextOverflow.None })
        .fontSize(12).border({ width: 1 }).padding(10)
      Text('这是一首画作欣赏诗,从诗中的描述来看,画中有山、水、花、鸟都是典型的中国画题材,而且肯定是一幅画得相当逼真、传神的作品。作者通过文字的描述,把一幅本是静止的画变成了一幅美丽的风景卷轴展现出来:苍翠的山,流动的水,绽放的花,欢鸣的鸟,一派鲜活的景象,把读者引入了无限的遐想之中。当读者从遐想中回到现实的时候,才发现,画中的一切不过是一个个无生命的静物。')
        .textOverflow({ overflow: TextOverflow.Clip })
        .maxLines(1).fontSize(12).border({ width: 1 }).padding(10)
      Text('这是一首画作欣赏诗,从诗中的描述来看,画中有山、水、花、鸟都是典型的中国画题材,而且肯定是一幅画得相当逼真、传神的作品。作者通过文字的描述,把一幅本是静止的画变成了一幅美丽的风景卷轴展现出来:苍翠的山,流动的水,绽放的花,欢鸣的鸟,一派鲜活的景象,把读者引入了无限的遐想之中。当读者从遐想中回到现实的时候,才发现,画中的一切不过是一个个无生命的静物。')
        .textOverflow({ overflow: TextOverflow.Ellipsis })
        .maxLines(1).fontSize(12).border({ width: 1 }).padding(10)

      Text('文本修饰示例').fontSize(19).fontColor(0xCCCCCC)
      Text('看远处的山往往是模糊的,但画上的山色却很清楚,在近处听流水,应当听到水声,但画上的流水却无声。')
        .decoration({ type: TextDecorationType.Underline, color: Color.Red })
        .fontSize(12).border({ width: 1 }).padding(10)
      Text('在春天盛开的花,随着春天的逝去就凋谢了。')
        .decoration({ type: TextDecorationType.LineThrough, color: Color.Red })
        .fontSize(12).border({ width: 1 }).padding(10)
      Text('画上的花,不管在什么季节,它都盛开着。')
        .decoration({ type: TextDecorationType.Overline, color: Color.Red })
        .fontSize(12).border({ width: 1 }).padding(10)
    }.height(600).width(350).padding({ left: 35, right: 35, top: 35 })
  }
}
```

### 3. 实验结果

在 DevEco Studio 3.0 中使用 Previewer 预览,得到如图 4.11 所示的 App 界面。图 4.11 中展示了文本的格式化,包括行高、文本溢出和文本下画线等。

### 4. 扩展

eTS 中通过调用接口 Text(content?: string)创建 Text 组件,其中,content 用于设置文本内容。Text 可以包含子组件 Span,在包含子组件 Span 时 content 无效,显示 Span 内容。

Text 组件的属性如下。

(1) textAlign 属性用于设置多行文本的文本对齐方式,其参数类型为枚举 TextAlign。
```

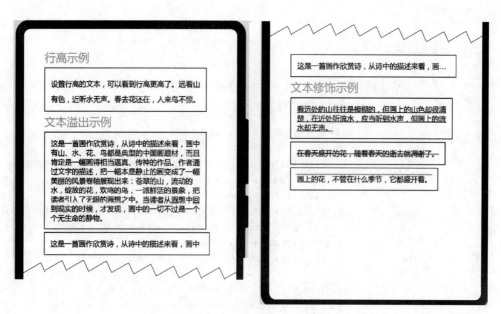

图 4.11　eTS 组件 Text 示例

该枚举包含 Center、Start 和 End 三个值。

（2）textOverflow 属性用于设置文本超长时的显示方式，其参数类型为 TextOverflow 枚举，该枚举包含 Clip（文本超长时进行裁剪显示）、Ellipsis（文本超长时显示不下的文本用省略号代替）和 None（文本超长时不进行裁剪）三个值。

（3）maxLines 属性用于设置文本的最大行数。

（4）lineHeight 属性用于设置文本的文本行高，设置值不大于 0 时，不限制文本行高，自适应字体大小，Length 为 number 类型时单位为 fp。

（5）baselineOffset 属性用于设置文本基线的偏移量。

（6）textCase 属性用于设置文本大小写，其参数类型为 TextCase 枚举，包含 Normal（保持文本原有大小写）、LowerCase（文本采用全小写）和 UpperCase（文本采用全大写）三个值。

（7）decorations 属性用于设置文本装饰线样式及其颜色。其参数为

{

　　type: TextDecorationType,

　　color?: Color

}

其中，TextDecorationType 枚举包括 Underline（文字下画线修饰）、LineThrough（穿过文本的修饰线）、Overline（文字上画线修饰）和 None（不使用文本修饰线）。

以下所列代码示例了文本大小写属性、对齐属性和位置偏移属性，将其内容替换 index.ets 的原有代码后，可通过 Previewer 预览效果。

```
@Entry
@Component
struct TextExample2 {
    build() {
        Flex({ direction: FlexDirection.Column, alignItems: ItemAlign.Start,
justifyContent: FlexAlign.SpaceBetween }) {
            Text('大小写示例').fontSize(19).fontColor(0xCCCCCC)
            Text('普通 text content with textCase set to Normal.')
                .textCase(TextCase.Normal)
                .fontSize(12).border({ width: 1 }).padding(10).width('100%')
            Text('小写 text content with textCase set to LowerCase.')
                .textCase(TextCase.LowerCase)
                .fontSize(12).border({ width: 1 }).padding(10).width('100%')
            Text('大写 text content with textCase set to UpperCase.')
                .textCase(TextCase.UpperCase)
                .fontSize(12).border({ width: 1 }).padding(10)

            Text('textAlign').fontSize(9).fontColor(0xCCCCCC)
            Text('居中对齐 text content with textAlign set to Center.')
                .textAlign(TextAlign.Center)
                .fontSize(12).border({ width: 1 }).padding(10).width('100%')
            Text('靠左对齐 text content with textAlign set to Start.')
                .textAlign(TextAlign.Start)
                .fontSize(12).border({ width: 1 }).padding(10).width('100%')
            Text('靠右对齐 text content with textAlign set to End.')
                .textAlign(TextAlign.End)
                .fontSize(12).border({ width: 1 }).padding(10).width('100%')

            Text('baselineOffset').fontSize(9).fontColor(0xCCCCCC)
            Text('位置偏移 text content with baselineOffset set to 10.')
                .baselineOffset(10).fontSize(12).border({ width: 1 }).padding(10).width('100%')
            Text('This is the text content with baselineOffset set to 30.')
                .baselineOffset(30).fontSize(12).border({ width: 1 }).padding(10).width('100%')
            Text('This is the text content with baselineOffset set to -10.')
                .baselineOffset(-10).fontSize(12).border({ width: 1 }).padding(10).width('100%')
        }.height(700).width(350).padding({ left: 35, right: 35, top: 35 })
    }
}
```

4.13 声明式开发 UI 组件 Image 实验

1. 实验内容

本实验练习声明式开发 UI 组件 Image。

2. 实验代码

在 DevEco Studio 3.0 中新建一个项目，在 Choose Your Ability Template 页选择

Empty Template，单击 Next 按钮进入 Config Your Project 页，在 Project name 框里填入项目名称，Project type 选择 Application，Development mode 选择 Traditional coding，Language 选择 eTS。单击 Finish 按钮。将以下代码替换掉该工程中 index.ets 文件中的内容。在目录 media 中创建三个图片文件，分别是 bird.gif、woman.png 和 example.jpg。

代码如下。

```
@Entry
@Component
struct ImageExample1 {
    private on: string = 'https://static.gushixuexi.com/wp-content/uploads/2016/11/2-43.jpg'
    @State src: string = this.on

    build() {
        Column() {
            Flex({ direction: FlexDirection.Column, alignItems: ItemAlign.Start }) {
                Text('default').fontSize(16).fontColor(0xcccccc).height(30)
                Row({ space: 5 }) {
                    Image($r('app.media.woman'))
                        .width(110).height(110).border({ width: 1 }).borderStyle(BorderStyle.Dashed)
                        .overlay('png', { align: Alignment.Bottom, offset: { x: 0, y: 20 } })

                    Image($r('app.media.bird'))
                        .width(110).height(110).border({ width: 1 }).borderStyle(BorderStyle.Dashed)
                        .overlay('gif', { align: Alignment.Bottom, offset: { x: 0, y: 20 } })

                    Image($r('app.media.woman'))
                        .width(110).height(110).border({ width: 1 }).borderStyle(BorderStyle.Dashed)
                        .overlay('svg', { align: Alignment.Bottom, offset: { x: 0, y: 20 } })
                }
                Row({ space: 5 }) {
                    Image($r('app.media.example'))
                        .width(110).height(110).border({ width: 1 }).borderStyle(BorderStyle.Dashed)
                        .overlay('jpg', { align: Alignment.Bottom, offset: { x: 0, y: 20 } })

                    Image(this.src)
                        .width(110).height(110).border({ width: 1 }).borderStyle(BorderStyle.Dashed)
                        .overlay('network', { align: Alignment.Bottom, offset: { x: 0, y: 20 } })
                }.margin({ top: 25, bottom: 10 })
            }
```

```
                            Flex({ direction: FlexDirection.Column, alignItems: ItemAlign.
Start }) {
                                Text('objectFit').fontSize(16).fontColor(0xcccccc).height
(30)
                                Row({ space: 5 }) {
                                    Image($r('app.media.example'))
                                        .border({ width: 1 }).borderStyle(BorderStyle.
Dashed)
                                        .objectFit(ImageFit.None).width(110).height
(110)
                                        .overlay('None', { align: Alignment.Bottom,
offset: { x: 0, y: 20 } })
                                    Image($r('app.media.example'))
                                        .border({ width: 1 }).borderStyle(BorderStyle.
Dashed)
                                        .objectFit(ImageFit.Fill).width(110).height
(110)
                                        .overlay('Fill', { align: Alignment.Bottom,
offset: { x: 0, y: 20 } })
                                    Image($r('app.media.example'))
                                        .border({ width: 1 }).borderStyle(BorderStyle.
Dashed)
                                        .objectFit(ImageFit.Cover).width(110).height
(110)
                                        .overlay('Cover', { align: Alignment.Bottom, offset:
{ x: 0, y: 20 } })
                                }
                                Row({ space: 5 }) {
                                    Image($r('app.media.example'))
                                        .border({ width: 1 }).borderStyle(BorderStyle.
Dashed)
                                        .objectFit(ImageFit.Contain).width(110).height
(110)
                                        .overlay('Contain', { align: Alignment.Bottom,
offset: { x: 0, y: 20 } })
                                    Image($r('app.media.example'))
                                        .border({ width: 1 }).borderStyle(BorderStyle.
Dashed)
                                        .objectFit(ImageFit.ScaleDown).width(110).height
(110)
                                        .overlay('ScaleDown', { align: Alignment.Bottom,
offset: { x: 0, y: 20 } })
                                }.margin({ top: 25 })
                            }
                        }.height(320).width(360).padding({ right: 10, top: 10 })
                    }
                }
```

3. 实验结果

在DevEco Studio 3.0中使用Previewer预览，得到如图4.12所示的App界面。

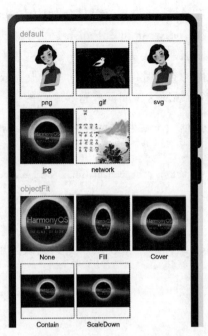

图 4.12　eTS 组件 Image 示例

图 4.12 展示了 png、gif 和 jpg 格式的图片,以及缩放类型等效果。

4. 扩展

Image 组件通过调用接口 Image(src: string | PixelMap)来创建。其中,src 参数可以为本地图片和网络图片的 URI,或者使用媒体 PixelMap 对象。

如果需要缩放图片,使用 objectFit 属性。该属性的参数为 ImageFit 枚举,其值有 Cover(保持宽高比进行缩小或者放大,使得图片两边都大于或等于显示边界)、Contain(保持宽高比进行缩小或者放大,使得图片完全显示在显示边界内)、Fill(不保持宽高比进行放大缩小,使得图片填充满显示边界)、None(保持原有尺寸显示,通常配合 objectRepeat 属性一起使用)和 ScaleDown(保持宽高比显示,图片缩小或者保持不变)。

如果设置图片解码尺寸,将原始图片解码成指定尺寸的图片,使用 sourceSize 属性,设置 width 和 height,单位为 px。

图片放大时,图片的插值效果通过 interpolation 属性设置。该属性的值为枚举 ImageInterpolation,包括 None(不使用插值图片数据)、High(高度使用插值图片数据,可能会影响图片渲染的速度)、Medium(中度使用插值图片数据)和 Low(低度使用插值图片数据)。

图片的渲染模式通过属性 renderMode 设置。该属性的值为 ImageRenderMode 枚举,包括 Original(按照原图进行渲染,包括颜色)和 Template(将图像渲染为模板图像,忽略图片的颜色信息)。

图片的重复样式通过属性 objectRepeat 设置。该属性的值为 ImageRepeat 枚举,包括 X(只在水平轴上重复绘制图片)、Y(只在竖直轴上重复绘制图片)、XY(在两个轴上重复绘制图片)和 NoRepeat(不重复绘制图片)。

加载时显示的占位图通过属性 alt 设置,该属性的值为 string 或 PixelMap。

Image 组件有 onComplete、onError 和 onFinish 三个事件。onComplete 事件在图片成功加载时触发,返回成功加载的图源尺寸。onError 事件在图片加载出现异常时触发该回调。加载的源文件为带动效的 svg 图片时,当 svg 动效播放完成时会触发 onFinish 回调,如果动效为无限循环动效,则不会触发这个回调。这些事件的接口如下。

```
onComplete(callback: (event?: { width: number, height: number, componentWidth: number, componentHeight: number, loadingStatus: number }) => void)

onError(callback: (event?: { componentWidth: number, componentHeight: number }) => void)

onFinish(callback: () => void)
```

以下的代码示例了 Image 组件的渲染模式、占位符、改变解码尺寸和重复模式属性的设置。

```
@Entry
@Component
struct ImageExample2 {
    @State width: number = 100
    @State height: number = 100

    build() {
        Column({ space: 10 }) {
            Text('渲染模式').fontSize(20).fontColor(0xcccccc).width('96%').height(30)
            Row({ space: 50 }) {
                Image($r('app.media.example'))
                    .renderMode(ImageRenderMode.Original).width(100).height(100)
                    .border({ width: 1 }).borderStyle(BorderStyle.Dashed)
                    .overlay('Original', { align: Alignment.Bottom, offset: { x: 0, y: 20 } })

                Image($r('app.media.example'))
                    .renderMode(ImageRenderMode.Template).width(100).height(100)
                    .border({ width: 1 }).borderStyle(BorderStyle.Dashed)
                    .overlay('Template', { align: Alignment.Bottom, offset: { x: 0, y: 20 } })
            }

            Text('占位符').fontSize(20).fontColor(0xcccccc).width('96%').height(30)
            Image('')
                .alt($r('app.media.example'))
                .width(100).height(100).border({ width: 1 }).borderStyle(BorderStyle.Dashed)

            Text('改变解码尺寸').fontSize(20).fontColor(0xcccccc).width('96%')
            Row({ space: 50 }) {
                Image($r('app.media.example'))
```

```
                .sourceSize({
                    width: 150,
                    height: 150
                })
                .objectFit(ImageFit.ScaleDown).width('25%').aspectRatio(1)
                .border({ width: 1 }).borderStyle(BorderStyle.Dashed)
                .overlay('w: 150 h: 150', { align: Alignment.Bottom, offset: {
x: 0, y: 20 } })
                Image($r('app.media.example'))
                .sourceSize({
                    width: 200,
                    height: 200
                })
                .objectFit(ImageFit.ScaleDown).width('25%').aspectRatio(1)
                .border({ width: 1 }).borderStyle(BorderStyle.Dashed)
                .overlay('w: 200 h: 200', { align: Alignment.Bottom, offset: {
x: 0, y: 20 } })
            }

            Text('重复模式').fontSize(20).fontColor(0xcccccc).width('96%').
height(30)

            Row({ space: 5 }) {
                Image($r('app.media.icon'))
                .width(120).height(125).border({ width: 1 }).borderStyle
(BorderStyle.Dashed)
                .objectRepeat(ImageRepeat.XY).objectFit(ImageFit.ScaleDown)
                .overlay('ImageRepeat.XY', { align: Alignment.Bottom, offset: {
x: 0, y: 20 } })
                Image($r('app.media.icon'))
                .width(110).height(125).border({ width: 1 }).borderStyle
(BorderStyle.Dashed)
                .objectRepeat(ImageRepeat.Y).objectFit(ImageFit.ScaleDown)
                .overlay('ImageRepeat.Y', { align: Alignment.Bottom, offset: {
x: 0, y: 20 } })
                Image($r('app.media.icon'))
                .width(110).height(125).border({ width: 1 }).borderStyle
(BorderStyle.Dashed)
                .objectRepeat(ImageRepeat.X).objectFit(ImageFit.ScaleDown)
                .overlay('ImageRepeat.X', { align: Alignment.Bottom, offset: {
x: 0, y: 20 } })
            }
        }.height(150).width('100%').padding({ right: 10 })
    }
}
```

该示例的界面显示如图 4.13 所示。

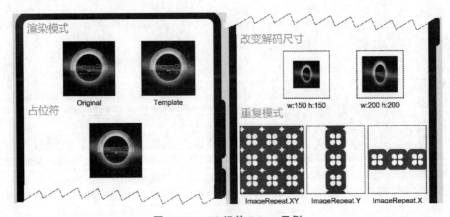

图 4.13　eTS 组件 Image 示例

以下的代码示例了 Image 组件的事件应用。

```
@Entry
@Component
struct ImageExample3 {
    @State width: number = 0
    @State height: number = 0
    private on: Resource = $r('app.media.wifi_on')
    private off: Resource = $r('app.media.wifi_off')
    private on2off: Resource = $r('app.media.wifi_on2off')
    private off2on: Resource = $r('app.media.wifi_off2on')
    @State src: Resource = this.on

    build() {
        Column() {
            Row({ space: 20 }) {
                Column() {
                    Image( $r('app.media.example'))
                    .alt( $r('app.media.ic_public_picture'))
                    .sourceSize({
                        width: 900,
                        height: 900
                    })
                    .objectFit(ImageFit.Cover)
                    .height(180).width(180)
                    .onComplete((msg: { width: number, height: number }) => {
                        this.width = msg.width
                        this.height = msg.height
                    })
                    .onError(() => {
                        console.log('load image fail')
                    })
                    .overlay('\nwidth: ' + String(this.width) + ' height: ' +
String(this.height), {
                        align: Alignment.Bottom,
                        offset: { x: 0, y: 20 }
```

```
                })
            }
            Image(this.src)
                .width(120).height(120)
                .onClick(() => {
                    if (this.src == this.on || this.src == this.off2on) {
                        this.src = this.on2off
                    } else {
                        this.src = this.off2on
                    }
                })
                .onFinish(() => {
                    if (this.src == this.off2on) {
                        this.src = this.on
                    } else {
                        this.src = this.off
                    }
                })
        }.width('100%')
    }
}
```

4.14 声明式开发 UI 组件 Slider 实验

1. 实验内容

本实验练习声明式开发 UI 组件 Slider。

2. 实验代码

在 DevEco Studio 3.0 中新建一个项目，在 Choose Your Ability Template 页选择 Empty Template，单击 Next 按钮进入 Config Your Project 页，在 Project name 框里填入项目名称，Project type 选择 Application，Development mode 选择 Traditional coding，Language 选择 eTS。单击 Finish 按钮。将以下代码替换掉该工程中 index.ets 文件中的内容。

代码如下。

```
@Entry
@Component
struct SliderExample{
    @State outSetValue: number = 40
    @State inSetValue: number = 40

    build() {
        Column({ space: 5 }) {
            Text('slider out set').fontSize(19).fontColor(0xCCCCCC).width('90%')
```

```
Row() {
    Slider({
        value: this.outSetValue,
        min: 0,
        max: 100,
        step: 1,
        style: SliderStyle.OutSet
    })
        .blockColor(Color.Blue)
        .trackColor(Color.Gray)
        .selectedColor(Color.Blue)
        .showSteps(true)
        .showTips(true)
        .onChange((value: number, mode: SliderChangeMode) => {
            this.outSetValue = value
            console.info('value: ' + value + 'mode: ' + mode.toString())
        })
    Text(this.outSetValue.toFixed(0)).fontSize(16)
}
.padding({ top: 50 })
.width('80%')

Text('slider in set').fontSize(19).fontColor(0xCCCCCC).width('90%')
Row() {
    Slider({
        value: this.inSetValue,
        min: 0,
        max: 100,
        step: 1,
        style: SliderStyle.InSet
    })
        .blockColor(0xCCCCCC)
        .trackColor(Color.Black)
        .selectedColor(0xCCCCCC)
        .showSteps(false)
        .showTips(false)
        .onChange((value: number, mode: SliderChangeMode) => {
            this.inSetValue = value
            console.info('value: ' + value + 'mode: ' + mode.toString())
        })
    Text(this.inSetValue.toFixed(0)).fontSize(16)
}
.width('80%')
}.width('100%').margin({ top: 5 })
    }
}
```

3. 实验结果

在 DevEco Studio 3.0 中使用 Previewer 预览，得到如图 4.14 所示的 App 界面。

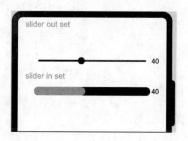

图 4.14 eTS 组件 Slider 示例

4. 扩展

Slider 组件通过调用接口 Slider(value：value?：number，min?：number，max?：number，step?：number，style?：SliderStyle)来创建。其中，value 参数表示当前进度值，min 参数设置最小值，max 参数设置最大值，step 设置 Slider 滑动跳动值，当设置相应的 step 时，Slider 为间歇滑动，style 参数设置 Slider 的滑块样式，值为 SliderStyle 枚举，包含 OutSet 和 InSet 两个值。

Slider 组件的滑块的颜色、滑轨的背景颜色和滑轨的已滑动颜色分别通过属性 blockColor、trackColor 和 selectedColor 设置。Slider 组件的事件有 OnAppear、OnDisAppear 和 onChange(callback：(value：number，mode：SliderChangeMode) => void)。

4.15 声明式开发 UI 组件 Flex 实验

1. 实验内容

本实验熟悉声明式开发 UI 容器组件 Flex。

2. 实验代码

在 DevEco Studio 3.0 中新建一个项目，在 Choose Your Ability Template 页选择 Empty Template，单击 Next 按钮进入 Config Your Project 页，在 Project name 框里填入项目名称，Project type 选择 Application，Development mode 选择 Traditional coding，Language 选择 eTS。单击 Finish 按钮。将以下代码替换掉该工程中 index.ets 文件中的内容。

代码如下。

```
@Entry
@Component
struct FlexExample1 {
    build() {
        Column() {
            Column({ space: 5 }) {
```

```
                    Text('横向顺排').fontSize(19).fontColor(0xCCCCCC).width('
90%')
                    Flex({ direction: FlexDirection.Row }) {
                        Text('子组件 1').fontSize(19).width('20%').
height(50).backgroundColor(0xF5DEB3)
                        Text('子组件 2').fontSize(19).width('20%').
height(50).backgroundColor(0xD2B48C)
                        Text('子组件 3').fontSize(19).width('20%').
height(50).backgroundColor(0xF5DEB3)
                        Text('子组件 4').fontSize(19).width('20%').
height(50).backgroundColor(0xD2B48C)
                    }
                    .height(70)
                    .width('90%')
                    .padding(10)
                    .backgroundColor(0xAFEEEE)

                    Text('横向逆排').fontSize(19).fontColor(0xCCCCCC).width('
90%')
                    Flex({ direction: FlexDirection.RowReverse }) {
                        Text('子组件 1').fontSize(19).width('20%').height
(50).backgroundColor(0xF5DEB3)
                        Text('子组件 2').fontSize(19).width('20%').height
(50).backgroundColor(0xD2B48C)
                        Text('子组件 3').fontSize(19).width('20%').height
(50).backgroundColor(0xF5DEB3)
                        Text('子组件 4').fontSize(19).width('20%').height
(50).backgroundColor(0xD2B48C)
                    }
                    .height(70)
                    .width('90%')
                    .padding(10)
                    .backgroundColor(0xAFEEEE)

                    Text('纵向顺排').fontSize(19).fontColor(0xCCCCCC).width('
90%')
                    Flex({ direction: FlexDirection.Column }) {
                        Text('子组件 1').fontSize(19).width('100%').height
(40).backgroundColor(0xF5DEB3)
                        Text('子组件 2').fontSize(19).width('100%').height
(40).backgroundColor(0xD2B48C)
                        Text('子组件 3').fontSize(19).width('100%').height
(40).backgroundColor(0xF5DEB3)
                        Text('子组件 4').fontSize(19).width('100%').height
(40).backgroundColor(0xD2B48C)
                    }
                    .height(160)
                    .width('90%')
                    .padding(10)
                    .backgroundColor(0xAFEEEE)
```

```
                    Text('纵向逆排').fontSize(19).fontColor(0xCCCCCC).width('
90%')
                        Flex({ direction: FlexDirection.ColumnReverse }) {
                            Text('子组件 1').fontSize(19).width('100%').height
(40).backgroundColor(0xF5DEB3)
                            Text('子组件 2').fontSize(19).width('100%').height
(40).backgroundColor(0xD2B48C)
                            Text('子组件 3').fontSize(19).width('100%').height
(40).backgroundColor(0xF5DEB3)
                            Text('子组件 4').fontSize(19).width('100%').height
(40).backgroundColor(0xD2B48C)
                        }
                        .height(160)
                        .width('90%')
                        .padding(10)
                        .backgroundColor(0xAFEEEE)
                }.width('100%').margin({ top: 5 })
            }.width('100%')
        }
    }
```

3. 实验结果

在 DevEco Studio 3.0 中使用 Previewer 预览,得到如图 4.15 所示的 App 界面。

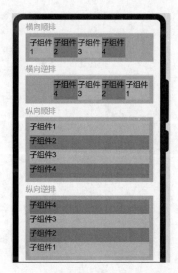

图 4.15　eTS 组件 Flex 子组件排列示例

4. 扩展

Flex 组件通过调用接口 Flex(options?: direction?: FlexDirection, wrap?: FlexWrap, justifyContent?: FlexAlign, alignItems?: ItemAlign, alignContent?: FlexAlign)来创建。其中,direction 参数表示子组件在 Flex 容器上排列的方向,wrap 参数设置 Flex 容器是单行/列还是多行/列排列,justifyContent 参数设置子组件在 Flex 容器主轴上的对齐格式,alignItems 参数设置子组件在 Flex 容器交叉轴上的对齐格式,alignContent 参数设置交叉

轴中有额外的空间时,多行内容的对齐方式。

参数 direction 的值为 FlexDirection 枚举,枚举项 Row 表示主轴与行方向一致作为布局模式,RowReverse 表示与 Row 方向相反方向进行布局,Column 表示主轴与列方向一致作为布局模式,ColumnReverse 表示与 Column 相反方向进行布局。

参数 wrap 的值为 FlexWrap。枚举项 NoWrap 表示 Flex 容器的子组件单行/列布局,子项允许超出容器。Wrap 表示 Flex 容器的子组件多行/列排布,子项允许超出容器。WrapReverse 表示 Flex 容器的子组件反向多行/列排布,子项允许超出容器。示例代码如下。

```
@Entry
@Component
struct FlexExample2{
    build() {
        Column() {
            Column({ space: 5 }) {
                Text('Wrap').fontSize(19).fontColor(0xCCCCCC).width('90%')
                Flex({ wrap: FlexWrap.Wrap }) {
                    Text('1').width('50%').height(50).backgroundColor(0xF5DEB3)
                    Text('2').width('50%').height(50).backgroundColor(0xD2B48C)
                    Text('3').width('50%').height(50).backgroundColor(0xD2B48C)
                }
                .width('90%')
                .padding(10)
                .backgroundColor(0xAFEEEE)

                Text('NoWrap').fontSize(19).fontColor(0xCCCCCC).width('90%')
                Flex({ wrap: FlexWrap.NoWrap }) {
                    Text('1').width('50%').height(50).backgroundColor(0xF5DEB3)
                    Text('2').width('50%').height(50).backgroundColor(0xD2B48C)
                    Text('3').width('50%').height(50).backgroundColor(0xF5DEB3)
                }
                .width('90%')
                .padding(10)
                .backgroundColor(0xAFEEEE)

                Text('WrapReverse').fontSize(19).fontColor(0xCCCCCC).width('90%')
                Flex({ wrap: FlexWrap.WrapReverse , direction: FlexDirection.Row }) {
                    Text('1').width('50%').height(50).backgroundColor(0xF5DEB3)
```

```
                                    Text('2').width('50%').height(50).
backgroundColor(0xD2B48C)
                                    Text('3').width('50%').height(50).
backgroundColor(0xD2B48C)
                                }
                                .width('90%')
                                .height(120)
                                .padding(10)
                                .backgroundColor(0xAFEEEE)
                        }.width('100%').margin({ top: 5 })
                }.width('100%')
        }
}
```

参数 justifyContent 的值为 FlexAlign，其枚举项如下。

（1）Start 表示子组件在主轴方向首端对齐，第一个子组件与行首对齐，同时后续的子组件与前一个对齐。

（2）Center 表示子组件在主轴方向中心对齐，第一个子组件与行首的距离与最后一个子组件与行尾的距离相同。

（3）End 表示子组件在主轴方向尾部对齐，最后一个子组件与行尾对齐，其他子组件与后一个对齐。

（4）SpaceBetween 表示 Flex 主轴方向均匀分配弹性子组件，相邻子组件之间距离相同。第一个子组件与行首对齐，最后一个子组件与行尾对齐。

（5）SpaceAround 表示 Flex 主轴方向均匀分配弹性子组件，相邻子组件之间距离相同。第一个子组件到行首的距离和最后一个子组件到行尾的距离是相邻子组件之间距离的一半。

（6）SpaceEvenly 表示 Flex 主轴方向子组件等间距布局，相邻子组件之间的间距、第一个子组件与行首的间距、最后一个子组件到行尾的间距都完全一样。

示例代码如下。

```
@Component
struct JustifyContentFlex{
        @Prop justifyContent : number

        build() {
                Flex({ justifyContent: this.justifyContent }) {
                        Text('1').width('33%').height(30).fontSize(25).
backgroundColor(0xF5DEB3)
                        Text('2').width('33%').height(40).fontSize(25).
backgroundColor(0xD2B48C)
                        Text('3').width('33%').height(50).fontSize(25).
backgroundColor(0xF5DEB3)
                }
                .width('90%')
                .padding(10)
                .backgroundColor(0xAFEEEE)
        }
```

```
            }
            @Entry
            @Component
            struct FlexExample3{
                    build() {
                            Column() {
                                    Column({ space: 5 }) {
                                            Text('justifyContent: Start').fontSize(19).fontColor(0xCCCCCC).width('90%')
                                            JustifyContentFlex({ justifyContent: FlexAlign.Start })

                                            Text('justifyContent: Center').fontSize(19).fontColor(0xCCCCCC).width('90%')
                                            JustifyContentFlex({ justifyContent: FlexAlign.Center })

                                            Text('justifyContent: End').fontSize(19).fontColor(0xCCCCCC).width('90%')
                                            JustifyContentFlex({ justifyContent: FlexAlign.End })

                                            Text('justifyContent: SpaceBetween').fontSize(19).fontColor(0xCCCCCC).width('90%')
                                            JustifyContentFlex({ justifyContent: FlexAlign.SpaceBetween })

                                            Text('justifyContent: SpaceAround').fontSize(19).fontColor(0xCCCCCC).width('90%')
                                            JustifyContentFlex({ justifyContent: FlexAlign.SpaceAround })

                                            Text('justifyContent: SpaceEvenly').fontSize(19).fontColor(0xCCCCCC).width('90%')
                                            JustifyContentFlex({ justifyContent: FlexAlign.SpaceEvenly })
                                    }.width('100%').margin({ top: 5 })
                            }.width('100%')
                    }
            }
```

该示例代码预览时的界面如图 4.16 所示。

参数 alignItems 的值为 ItemAlign,其枚举项如下。

(1) Auto 表示使用 Flex 容器中默认配置。

(2) Start 表示子组件在 Flex 容器中,交叉轴方向首部对齐。

(3) Center 表示子组件在 Flex 容器中,交叉轴方向居中对齐。

(4) End 表示子组件在 Flex 容器中,交叉轴方向尾部对齐。

(5) Stretch 表示子组件在 Flex 容器中,交叉轴方向拉伸填充,在未设置尺寸时,拉伸到容器尺寸。

(6) Baseline 表示子组件在 Flex 容器中,交叉轴方向文本基线对齐。

示例代码如下。

```
@Component
```

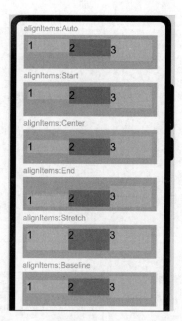

图 4.16 eTS 组件 Flex 子组件对齐示例

```
struct AlignItemsFlex{
    @Prop alignItems : number

    build() {
        Flex({ alignItems: this.alignItems }) {
            Text('1').fontSize(19).width('33%').height(30).backgroundColor(0xF5DEB3)
            Text('2').fontSize(19).width('33%').height(40).backgroundColor(0xD2B48C)
            Text('3').fontSize(19).width('33%').height(50).backgroundColor(0xF5DEB3)
        }
        .size({width: '90%', height: 80})
        .padding(10)
        .backgroundColor(0xAFEEEE)
    }
}

@Entry
@Component
struct FlexExample4{
    build() {
        Column() {
            Column({ space: 5 }) {
                Text('alignItems: Auto').fontSize(19).fontColor(0xCCCCCC).width('90%')
                AlignItemsFlex({ alignItems: ItemAlign.Auto })
```

```
                                    Text('alignItems: Start').fontSize(19).fontColor
(0xCCCCCC).width('90%')
                                    AlignItemsFlex({ alignItems: ItemAlign.Start })

                                    Text('alignItems: Center').fontSize(19).fontColor
(0xCCCCCC).width('90%')
                                    AlignItemsFlex({ alignItems: ItemAlign.Center })

                                    Text('alignItems: End').fontSize(19).fontColor
(0xCCCCCC).width('90%')
                                    AlignItemsFlex({ alignItems: ItemAlign.End })

                                    Text('alignItems: Stretch').fontSize(19).fontColor
(0xCCCCCC).width('90%')
                                    AlignItemsFlex({ alignItems: ItemAlign.Stretch })

                                    Text('alignItems: Baseline').fontSize(19).fontColor
(0xCCCCCC).width('90%')
                                    AlignItemsFlex({ alignItems: ItemAlign.Baseline })
                        }.width('100%').margin({ top: 5 })
            }.width('100%')
        }
    }
```

该示例代码预览时的界面如图 4.17 所示。

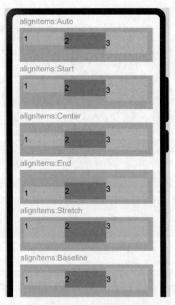

图 4.17　eTS 组件 Flex 子组件交叉轴对齐示例

4.16 声明式开发 UI 组件 Stack 实验

1. 实验内容

本实验熟悉声明式开发 UI 容器组件 Stack。Stack 是堆叠容器，子组件按照顺序依次入栈，后一个子组件覆盖前一个子组件。

2. 实验代码

在 DevEco Studio 3.0 中新建一个项目，在 Choose Your Ability Template 页选择 Empty Template，单击 Next 按钮进入 Config Your Project 页，在 Project name 框里填入项目名称，Project type 选择 Application，Development mode 选择 Traditional coding，Language 选择 eTS。单击 Finish 按钮。将以下代码替换掉该工程中 index.ets 文件中的内容。

组件 Stack 通过 Stack(options?：alignContent?：Alignment)接口创建。Alignment 枚举包含 TopStart、Top、TopEnd、Start、Center、End、BottomStart、Bottom 和 BottomEnd。

代码如下：

```
@Entry
@Component
struct StackExample{
    build() {
        Stack({ alignContent: Alignment.Bottom }) {
            Text('第一个子组件').width('90%').height('100%').backgroundColor(0xd2cab3).align(Alignment.Top)
            Text('第二个子组件').width('70%').height('80%').backgroundColor(0xc1cbac).align(Alignment.Top)
            Text('第三个子组件').width('50%').height('60%').backgroundColor(0xa1cbac).align(Alignment.Top)
        }.width('100%').height(150).margin({ top: 5 })
    }
}
```

3. 实验结果

在 DevEco Studio 3.0 中使用 Previewer 预览，得到如图 4.18 所示的 App 界面。

图 4.18 eTS 组件 Stack 示例

4.17 声明式开发 UI 组件 Tabs

1. 实验内容

本实验熟悉声明式开发 UI 容器组件 Tabs。该组件通过 Tab 进行内容视图的切换，每个 Tab 对应一个内容视图。

2. 实验代码

在 DevEco Studio 3.0 中新建一个项目，在 Choose Your Ability Template 页选择 Empty Template，单击 Next 按钮进入 Config Your Project 页，在 Project name 框里填入项目名称，Project type 选择 Application，Development mode 选择 Traditional coding，Language 选择 eTS。单击 Finish 按钮。将以下代码替换掉该工程中 index.ets 文件中的内容。

代码如下。

```
@Entry
@Component
struct TabsExample{
    private controller: TabsController = new TabsController()

    build() {
        Column() {
            Tabs({ barPosition: BarPosition.Start, index: 1, controller: this.controller }) {
                TabContent() {
                    Text('三字经').width('100%').height('100%').backgroundColor(Color.Pink)
                }.tabBar('语文')

                TabContent() {
                    Text('二项式分布').width('100%').height('100%').backgroundColor(Color.Yellow)
                }.tabBar('数学')

                TabContent() {
                    Text('俄乌战争').width('100%').height('100%').backgroundColor(Color.Blue)
                }.tabBar('政治')

                TabContent() {
                    Text('COVID-19').width('100%').height('100%').backgroundColor(Color.Green)
                }.tabBar('英语')
            }
            .vertical(true).scrollable(true).barMode(BarMode.Fixed)
            .barWidth(70).barHeight(150).animationDuration(400)
```

```
                    .onChange((index: number) => {
                        console.info(index.toString())
                    })
                    .width('90%').backgroundColor(0xF5F5F5)
            }.width('100%').height(150).margin({ top: 5 })
        }
    }
}
```

3. 实验结果

在 DevEco Studio 3.0 中使用 Previewer 预览,得到如图 4.19 所示的 App 界面。

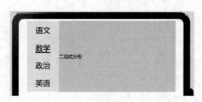

图 4.19　eTS 组件 Tabs 示例

4. 扩展

组件 Tabs 通过 Tabs（value：barPosition?：BarPosition, index?：number, controller?：TabsController）接口创建。参数 barPosition 指定 Tabs 位置来创建 Tabs 容器组件、参数 index 指定初次初始 Tab 索引、参数 controller 设置 Tabs 控制器。

参数 barPosition 的类型为 BarPosition 枚举。值为 Start 时,并且 vertical 属性方法设置为 true 时,Tab 标签位于容器左侧;vertical 属性方法设置为 false 时,Tab 标签位于容器顶部。值为 End,且 vertical 属性方法设置为 true 时,Tab 标签位于容器右侧;vertical 属性方法设置为 false 时,Tab 标签位于容器底部。

TabsController 为 Tabs 组件的控制器,用于控制 Tabs 组件进行页签切换。

该组件的属性如下。

(1) 属性 vertical 表示 Tab 是否为纵向排列,默认为 false。

(2) 属性 scrollable 表示是否可以通过左右滑动进行页面切换,默认为 true。

(3) 属性 barMode 指明 TabBar 布局模式,该属性类型为 BarMode 枚举,包含 Scrollable(TabBar 使用实际布局宽度,超过总长度后可滑动)和 Fixed(所有 TabBar 平均分配宽度)。

(4) 属性 barWidth 指明 TabBar 的宽度值,不设置时使用系统主题中的默认值。

(5) 属性 barHeight 指明 TabBar 的高度值,不设置时使用系统主题中的默认值。

(6) 属性 animationDuration 表示 TabContent 滑动动画时长,默认为 200。

该组件的事件有 Tab 页切换后触发的事件,接口为 onChange(callback：(index：number) => void)。

在 Tabs 中使用的组件为 TabContent,对应一个 Tab 的内容视图,通过 TabContent() 接口创建。该组件属性为 tabBar,参数类型为 string 或 icon?：string, text?：string。

4.18 声明式开发 UI 组件 List 实验

1. 实验内容

本实验熟悉声明式开发 UI 容器组件 List。该组件包含一系列相同宽度的列表项，适合连续、多行呈现同类数据，例如图片和文本。

2. 实验代码

在 DevEco Studio 3.0 中新建一个项目，在 Choose Your Ability Template 页选择 Empty Template，单击 Next 按钮进入 Config Your Project 页，在 Project name 框里填入项目名称，Project type 选择 Application，Development mode 选择 Traditional coding，Language 选择 eTS。单击 Finish 按钮。将以下代码替换掉该工程中 index.ets 文件中的内容。

代码如下。

```
@Entry
@Component
struct ListExample{
    private arr: string[] = ['太阳', '水星', '火星', '地球', '金星', '木星', '土星', '海王星', '冥王星', '哈雷彗星']
    @State editFlag: boolean = false

    build() {
        Stack({ alignContent: Alignment.TopStart }) {
            Column() {
                List({ space: 20, initialIndex: 0 }) {
                    ForEach(this.arr, (item) => {
                        ListItem() {
                            Text('' + item)
                                .width('100%').height(50).fontSize(25)
                                .textAlign(TextAlign.Center).borderRadius(10).backgroundColor(0xFFFFFF)
                        }.editable(true)
                    }, item => item)
                }
                .listDirection(Axis.Vertical)  //排列方向
                .divider({ strokeWidth: 2, color: 0xFFFFFF,
                    startMargin: 20, endMargin: 20 })  //每行之间的分界线
                .edgeEffect(EdgeEffect.None)  //滑动到边缘无效果
                .chainAnimation(false)  //联动特效关闭
                .onScrollIndex((firstIndex: number, lastIndex: number) => {
                    console.info('first' + firstIndex)
                    console.info('last' + lastIndex)
                })
                .editMode(this.editFlag)
```

```
                    .onItemDelete((index: number) => {
                            console.info(this.arr[index] + 'Delete')
                            this.arr.splice(index, 1)
                            console.info(JSON.stringify(this.arr))
                            this.editFlag = false
                            return true
                    }).width('90%')
            }.width('100%')

            Button('编辑列表')
            .onClick(() => {
                    this.editFlag = !this.editFlag
            }).margin({ top: 5, left: 20 })
    }.width('100%').height('100%').backgroundColor(0xDCDCDC).padding
({ top: 5 })
        }
    }
```

3. 实验结果

在 DevEco Studio 3.0 中使用 Previewer 预览，得到如图 4.20 所示的 App 界面。

图 4.20　eTS 组件 List 示例

4. 扩展

组件 List 通过 List(options?：space?：number，initialIndex?：number)接口创建。参数 space 指定列表项间距；参数 initialIndex 设置当前 List 初次加载时视口起始位置显示的 item，即显示第一个 item，如设置的序号超过了最后一个 item 的序号，则设置不生效。

该组件的属性如下。

（1）属性 listDirection 表示 List 组件排列方向参照 Axis 枚举说明，该枚举的值 Vertical 表示纵向排列，Horizontal 表示横向排列。

（2）属性 divider 用于设置 ListItem 分隔线样式，默认无分隔线。分隔线样式 strokeWidth：Length，color?：Color，startMargin?：Length，endMargin?：Length。strokeWidth 为分隔线的线宽，color 为分隔线的颜色，startMargin 为分隔线距离列表侧边

起始端的距离，endMargin 为分隔线距离列表侧边结束端的距离。

（3）属性 editMode 指明当前 List 组件是否处于可编辑模式，默认值为 false。

（4）属性 edgeEffect 指明滑动效果，目前支持的滑动效果通过枚举 EdgeEffect 设置。枚举 EdgeEffect 的值如果为 Spring，指弹性物理动效，滑动到边缘后可以根据初始速度或通过触摸事件继续滑动一段距离，松手后回弹。

（5）属性 chainAnimation 指明当前 List 是否启用链式联动动效，开启后列表滑动以及顶部和底部拖曳时会有链式联动的效果。链式联动效果是指组件 List 内的列表项间隔一定距离，在基本的滑动交互行为下，主动对象驱动从动对象进行联动，驱动效果遵循弹性物理动效。

该组件的事件有列表项删除时触发事件 onItemDelete((index: number) => boolean)和当前列表显示的起始位置和终止位置发生变化时触发事件 onScrollIndex((firstIndex: number, lastIndex: number) => void)。

在 List 容器组件中，ListItem 用来展示列表的具体项，通过 ListItem()创建，宽度默认充满 List 组件。ListItem 组件的 sticky 属性设置 ListItem 吸顶效果，其值为 Sticky 枚举，枚举值如为 Normal，表示当前 item 吸顶，滑动消失。ListItem 组件的属性 editable 声明当前 ListItem 元素是否可编辑，进入编辑模式后可删除，默认为 false。

4.19 声明式开发 UI 组件 Grid 实验

1. 实验内容

本实验熟悉声明式开发 UI 容器组件 Grid。该组件为网格容器，采用二维布局，将容器划分成"行"和"列"，产生单元格，然后指定"项目所在"的单元格，可以任意组合不同的网格，做出各种各样的布局。

2. 实验代码

在 DevEco Studio 3.0 中新建一个项目，在 Choose Your Ability Template 页选择 Empty Template，单击 Next 按钮进入 Config Your Project 页，在 Project name 框里填入项目名称，Project type 选择 Application，Development mode 选择 Traditional coding，Language 选择 eTS。单击 Finish 按钮。将以下代码替换掉该工程中 index.ets 文件中的内容。

代码如下。

```
@Entry
@Component
struct GridExample{
    @State Number: String[] = ['金', '木', '水', '火', '土']

    build() {
        Column({ space: 5 }) {
            Grid() {
                ForEach(this.Number, (day: string) => {
                    ForEach(this.Number, (day: string) => {
```

```
                        GridItem() {
                            Text(day)
                                .fontSize(16)
                                .backgroundColor(0xF9CF93)
                                .width('100 %')
                                .height('100 %')
                                .textAlign ( TextAlign.Center)
                        }
                    }, day => day)
                }, day => day)
            }
            .columnsTemplate('1fr 1fr 1fr 1fr 1fr')
            .rowsTemplate('1fr 1fr 1fr 1fr 1fr')
            .columnsGap(10)
            .rowsGap(10)
            .width('90 %')
            .backgroundColor(0xFAEEE0)
            .height(300)

            Text('scroll').fontColor(0xCCCCCC).fontSize(19).width('90 %')

            Grid() {
                ForEach(this.Number, (day: string) => {
                    ForEach(this.Number, (day: string) => {
                        GridItem() {
                            Text(day)
                                .fontSize(16)
                                .backgroundColor(0xF9CF93)
                                .width('100 %')
                                .height(80)
                                .textAlign ( TextAlign.Center)
                        }
                    }, day => day)
                }, day => day)
            }
            .columnsTemplate('1fr 1fr 1fr 1fr 1fr')
            .columnsGap(10)
            .rowsGap(10)
            .onScrollIndex((first: number) => {
                console.info(first.toString())
            })
            .width('90 %')
            .backgroundColor(0xFAEEE0)
            .height(300)
        }.width('100 %').margin({ top: 5 })
    }
}
```

3. 实验结果

在 DevEco Studio 3.0 中使用 Previewer 预览，得到如图 4.21 所示的 App 界面。

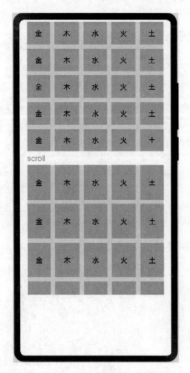

图 4.21　eTS 组件 Grid 示例

4. 扩展

组件 Grid 通过 Grid()接口创建，该组件的属性如下。

(1) 属性 columnsTemplate 用于设置网格布局的列，不设置时默认 1 列。该属性的参数类型为字符串，如'1fr 1fr 2fr'表示分为 3 列，将父组件宽度分为 4 等份，第 1 列占 1 份，第 2 列占 1 份，第 3 列占 2 份。

(2) 属性 rowsTemplate 用于设置网格布局的行，不设置时默认 1 行。该属性的参数类型为字符串，如'1fr 1fr 2fr'表示分为三行，将父组件高度分为 4 等份，第 1 行占 1 份，第 2 行占一份，第 3 行占 2 份。

(3) 属性 columnsGap 用于设置列与列的间距。

(4) 属性 rowsGap 用于设置行与行的间距。

当前列表显示的起始位置 item 发生变化时触发事件 onScrollIndex((first: number) => void)。

在 Grid 容器组件中，GridItem 为网格容器中单项内容容器，通过 GridItem()创建。有 rowStart、rowEnd、columnStart、columnEnd 和 forceRebuild 等属性。前四个属性用于指定 GridItem 起始行号、终点行号、起始列号和终点列号。forceRebuild 属性用于设置在触发组件 build 时是否重新创建此节点。示例代码如下。

@Entry

```
@Component
struct GridItemExample{
    @State numbers: string[] = Array.apply(null, Array(16)).map(function (item, i) { return i.toString() })

    build() {
        Column() {
            Grid() {
                GridItem() {
                    Text('占 4 行')
                        .fontSize(16).backgroundColor(0xFAEEE0)
                        .width('100%').height('100%').textAlign(TextAlign.Center)
                }.rowStart(1).rowEnd(4)

                ForEach(this.numbers, (item) => {
                    GridItem() {
                        Text(item)
                            .fontSize(16).backgroundColor(0xF9CF93)
                            .width('100%').height('100%').textAlign(TextAlign.Center)
                    }.forceRebuild(false)
                }, item => item)

                GridItem() {
                    Text('占 5 列')
                        .fontSize(16).backgroundColor(0xDBD0C0)
                        .width('100%').height('100%').textAlign(TextAlign.Center)
                }.columnStart(1).columnEnd(5)
            }
            .columnsTemplate('1fr 1fr 1fr 1fr 1fr')
            .rowsTemplate('1fr 1fr 1fr 1fr 1fr')
            .width('90%').height(300)
        }.width('100%').margin({ top: 5 })
    }
}
```

该示例的结果为如图 4.22 所示的 App 界面。

图 4.22 eTS 组件 Grid 示例

4.20 声明式开发 UI 自定义组件实验

1. 实验内容

本实验通过容器组件 Stack、Flex 和基本组件 Image、Text，构建用户自定义组件，完成图文并茂的人物介绍 App。

2. 实验代码

在 DevEco Studio 3.0 中新建一个项目，在 Choose Your Ability Template 页选择 Empty Template，单击 Next 按钮进入 Config Your Project 页，在 Project name 框里填入项目名称，Project type 选择 Application，Development mode 选择 Traditional coding，Language 选择 eTS。单击 Finish 按钮。

删掉工程模板的 build() 方法的代码，创建 Stack 组件，Stack 组件为堆叠组件，可以包含一个或多个子组件，其特点是后一个子组件覆盖前一个子组件。将 Text 组件放进 Stack 组件的花括号中，使其成为 Stack 组件的子组件。

创建 Image 组件，指定 Image 组件的 URL，Image 组件和 Text 组件都是必选构造参数组件。为了让 Text 组件在 Image 组件上方显示，所以要先声明 Image 组件。图片资源放在 resources 下的 rawfile 目录内，引用 rawfile 下资源时使用 $rawfile('filename')的形式，filename 为 rawfile 目录下的文件相对路径。当前 $rawfile 仅支持 Image 控件引用图片资源。

image 的 objectFit 属性默认为 ImageFit.Cover，设置为 ImageFit.Contain。可以设置高度 height 属性和宽度 width 属性。

Stack 默认为居中对齐，设置 Stack 构造参数 alignContent 为 Alignment.BottomStart，对齐方式为底部起始端对齐。

设置 Stack 的背景颜色来改变人物图片的背景颜色，通过框架提供的 Color 内置枚举值来设置，如 backgroundColor(Color.Red)，即设置背景颜色为红色。或者使用 string 类型参数，支持的颜色格式有 rgb、rgba 和 HEX 颜色码。如 backgroundColor('#0000FF')，即设置背景颜色为蓝色；backgroundColor('rgb(255,255,255)')，即设置背景颜色为白色。

调整 Text 组件的外边距 margin，使其距离左侧和底部有一定的距离。margin 是简写属性，可以统一指定四个边的外边距，也可以分别指定。参数为 Length 时，即统一指定四个边的外边距，如 margin(20)，即上、右、下、左四个边的外边距都是 20。参数为 top?: Length, right?: Length, bottom?: Length, left?: Length，即分别指定四个边的边距，如 margin(left: 20, bottom: 20)，即左边距为 20，下边距为 20。

核心代码如下。

```
@Entry
@Component
struct MyComponent {
    build() {
        Stack({ alignContent: Alignment.BottomStart }) {
            Image($r('app.media.Tomato'))
```

```
                .objectFit(ImageFit.Contain)
                .height(357)
                Text('Tomato')
                    .fontSize(26)
                    .fontWeight(500)
                    .margin({left: 26, bottom: 17.4})
            }
            .backgroundColor('#FFedf2f5')
    }
}
```

调整组件间的结构,语义化组件名称。创建页面入口组件为 BioDetail,在 BioDetail 中创建 Column,设置水平方向上居中对齐 alignItems(HorizontalAlign.Center)。Index 组件名改为 peopleImageDisplay,为 BioDetail 的子组件。Column 是子组件竖直排列的容器组件,本质为线性布局,所以只能设置交叉轴方向的对齐。

```
@Component
struct peopleImageDisplay {
    build() {
        Stack({ alignContent: Alignment.BottomStart }) {
            //Image($rawfile('zuchongzhi.PNG'))
            Image($r('app.media.zuchongzhi'))
                .objectFit(ImageFit.Contain)
                .height(300)
                .width(300)
            Text('祖冲之')
                .fontSize(50)
                .fontWeight(FontWeight.Bolder)
                .margin(20)
        }
        .backgroundColor(Color.Green)
    }
}

@Entry
@Component
struct BioDetail {
    build() {
        Column() {
            peopleImageDisplay()
        }
        .alignItems(HorizontalAlign.Center)
    }
}
```

可以使用 Flex 弹性布局来构建人物的简历表,Flex 通过比例来设置不同单元格的大小。创建 Flex 组件展示人物生卒和成就两部分。创建 Flex 组件,高度为 280,上、右、左内边距为 30,包含三个 Text 子组件分别代表姓名、出生年和死亡年。Flex 组件默认为水平排列方式。调整布局,设置各部分占比。姓名占比(layoutWeight)为 1,生卒年一共占比

(layoutWeight)为 2。出生年占据所有剩余空间 flexGrow(1)。成就这部分和生卒类似。

设置外层 Flex 为竖直排列 FlexDirection.Column,在主轴方向(竖直方向)上等距排列 FlexAlign.SpaceBetween,在交叉轴方向(水平轴方向)上首部对齐排列 ItemAlign.Start。

完整代码如下。

```
@Component
struct peopleImageDisplay {
    build() {
        Stack({ alignContent: Alignment.BottomStart }) {
            //Image( $ rawfile('zuchongzhi.PNG'))
            Image( $ r('app.media.zuchongzhi'))
                .objectFit(ImageFit.Contain)
                .height(300)
                .width(300)
            Text('祖冲之')
                .fontSize(50)
                .fontWeight(FontWeight.Bolder)
                .margin(20)
        }
        .backgroundColor(Color.Green)
    }
}

@Component
struct ContentTable {
    build() {
        Flex({ direction: FlexDirection.Column, justifyContent: FlexAlign.SpaceBetween, alignItems: ItemAlign.Start }) {
            Flex() {
                Text('祖冲之')
                    .fontSize(17.4)
                    .fontWeight(FontWeight.Bold)
                    .layoutWeight(1)
                Flex() {
                    Text('429 年')
                        .fontSize(17.4)
                        .flexGrow(1)
                    Text(' - 500 年')
                        .fontSize(17.4)
                }
                .layoutWeight(2)
            }
            Flex() {
                Text('成就')
                    .fontSize(17.4)
                    .fontWeight(FontWeight.Bold)
                    .layoutWeight(1)
                Flex() {
                    Text('数学')
```

```
                                    .fontSize(17.4)
                                    .flexGrow(1)
                                Text('7位圆周率')
                                    .fontSize(17.4)
                            }
                            .layoutWeight(2)
                        }

                        Flex() {
                            Text(' ')
                                .fontSize(17.4)
                                .fontWeight(FontWeight.Bold)
                                .layoutWeight(1)
                            Flex() {
                                Text('天文学')
                                    .fontSize(17.4)
                                    .flexGrow(1)
                                Text('大明历')
                                    .fontSize(17.4)
                            }
                            .layoutWeight(2)
                        }

                    }
                    .height(280)
                    .padding({ top: 30, right: 30, left: 30 })
                }
            }


@Entry
@Component
struct BioDetail {
    build() {
        Column() {
            peopleImageDisplay()
            ContentTable()
        }
        .alignItems(HorizontalAlign.Center)
    }
}
```

当前对每个成就都进行了声明,代码重复。可以使用@Builder来构建自定义方法,抽象出相同的UI结构声明。@Builder修饰的方法和Component的build方法都是为了声明一些UI渲染结构,遵循一样的eTS语法。可以定义一个或者多个@Builder修饰的方法,但Component的build方法必须只有一个。

在ContentTable内声明@Builder修饰的Item方法,用于声明成就、学科和贡献的UI描述。在ContentTable的build方法内调用Item接口,需要用this去调用该Component作用域内的方法,以此来区分全局的方法调用。ContentTable组件整体代码如下。

```
@Component
struct ContentTable {
    @Builder IngredientItem(title: string, name: string, value: string) {
        Flex() {
            Text(title)
                .fontSize(17.4)
                .fontWeight(FontWeight.Bold)
                .layoutWeight(1)
            Flex() {
                Text(name)
                    .fontSize(17.4)
                    .flexGrow(1)
                Text(value)
                    .fontSize(17.4)
            }
            .layoutWeight(2)
        }
    }

    build() {
        Flex({ direction: FlexDirection.Column, justifyContent: FlexAlign.SpaceBetween, alignItems: ItemAlign.Start }) {
            this.Item('祖冲之','429 年','- 500 年')
            this.Item('成就','数学','7 位圆周率')
            this.Item(' ','天文学','大明历')
            this.Item(' ','农业','水碓磨')
        }
        .height(280)
        .padding({ top: 30, right: 30, left: 30 })
    }
}
```

3. 实验结果

本实验 App 的界面如图 4.23 所示。

图 4.23　历史人物祖冲之的 App 界面

4.21 声明式开发多组件 UI 实验

1. 实验内容

在本实验中，构建了地区列表页面和地区气象参数详情页，展示了 List 布局、Grid 布局以及页面路由的基本用法。

2. 数据模型

地区气象的各个信息，如地理位置、全年最高温度、全年最低温度、全年降水量和全年空气质量等，要创建数据模型来统一存储和管理数据。如北京，东经 116°20′，北纬 39°56′，夏季最高温度 35℃、冬季最低温度-12℃、全年降水量 483.9mm 和全年空气质量中等。

(1) 新建 model 目录，在 model 目录下创建 climateData.ets。

(2) 定义气象参数数据的存储模型 climateData 和枚举变量 Category，climateData 类包含地区 id、名称(name)、分类(category)、图片(image)、经度(longitude)、纬度(latitude)、最高气温(temceiling)、最低气温(tembottom)、降水量(precipitation)和空气质量(airquality)属性。

```
//climateData.ets
enum Category {
    //一线城市,新一线城市、二线城市、三线城市、四线城市、五线城市
    //此例添加北京、武汉、大连、三亚、安阳、玉树 6 个城市
    cityclass0,
    cityclass1,
    cityclass2,
    cityclass3,
    cityclass4,
    cityclass5
}

let NextId = 0;
class climateData {
    id: string;
    name: string;
    image: Resource;
    category: Category;
    longitude: number;
    latitude: number;
    temceiling: number;
    tembottom: number;
    precipitation: number;
    airquality: number;

    constructor(name: string, image: Resource, category: Category,
    longitude: number, latitude: number, temceiling: number,
    tembottom: number, precipitation: number, airquality: number) {
        this.id = `${ NextId++}`;
        this.name = name;
```

```
            this.image = image;
            this.category = category;
            this.longitude = longitude;
            this.latitude = latitude;
            this.temceiling = temceiling;
            this.tembottom = tembottom;
            this.precipitation = precipitation;
            this.airquality = airquality;
        }
}
```

（3）在 resources→phone→media 目录下存入地区图片资源，图片名称为地区名称。这里添加了北京、武汉、大连、三亚、安阳和玉树 6 个城市。

（4）创建气候资源数据。在 model 目录下创建 climateDataModels.ets，在该页面中声明气候参数数组 climateComposition。

（5）创建 initializeOnStartUp 方法来初始化 climateData 的数组。在 climateDataModels.ets 中使用了定义在 climateData.ets 的 climateData 和 Category，所以要将 climateData.ets 的 climateData 类导出，在 climateDataModels.ets 内导入 climateData 和 Category。

```
//climateDataModels.ets
          import { Category, climateData } from './climateData.ets'

          const climateComposition: any[] = [
            { 'name': '北京', 'image': $r('app.media.Beijing'), 'category': Category.cityclass0,
'longitude': 17, 'latitude': 0.9, 'temceiling': 35, 'tembottom': -16, 'precipitation': 17.8,
'airquality': 123 },
            { 'name': '武汉', 'image': $r('app.media.Wuhan'), 'category': Category.cityclass1,
'longitude': 654, 'latitude': 15, 'temceiling': 41, 'tembottom': -5, 'precipitation': 1.3,
'airquality': 150 },
            { 'name': '大连', 'image': $r('app.media.Dalian'), 'category': Category.cityclass2,
'longitude': 30, 'latitude': 3, 'temceiling': 34, 'tembottom': -1, 'precipitation': 2.1,
'airquality': 23 },
            { 'name': '三亚', 'image': $r('app.media.Sanya'), 'category': Category.cityclass3,
'longitude': 57, 'latitude': 0.7, 'temceiling': 38, 'tembottom': 9, 'precipitation': 9.7,
'airquality': 12 },
            { 'name': '安阳', 'image': $r('app.media.Anyang'), 'category': Category.cityclass4,
'longitude': 97, 'latitude': 19, 'temceiling': 36, 'tembottom': -10, 'precipitation': 7.6,
'airquality': 223 },
            { 'name': '玉树', 'image': $r('app.media.Yushu'), 'category': Category.cityclass5,
'longitude': 207, 'latitude': 3.5, 'temceiling': 32, 'tembottom': -15, 'precipitation': 0.6,
'airquality': 43 }
          ]

          export functioninitializeOnStartup(): Array<climateData> {
            let climateDataArray: Array<climateData> = []
            limateComposition.forEach(item => {
                climateDataArray.push(new climateData(item.name, item.image, item.
category, item.longitude, item.latitude, item.temceiling, item.tembottom, item.
```

```
                precipitation, item.airquality ));
            })
            return climateDataArray;
    }
```

3. 列表组件 List 布局

List 组件是列表组件，适用于重复同类数据的展示，其子组件为 ListItem，适用于展示列表中的单元。使用 List 组件和 ForEach 循环渲染，构建地区列表布局，其 App 界面如图 4.24 所示。

图 4.24　地区列表布局

（1）在 pages 目录新建页面 climateAreaList.ets、climateDetail.ets，并将其添加到 config.json 文件下的 pages 标签，位于第一序位的页面为首页。

```
"js": [
{
    "mode": {
        "syntax": "ets",
        "type": "pageAbility"
    },
    "pages": [
    "pages/climateAreaList.ets",
    "pages/climateDetail.ets",
    /"pages/index"
    ],
    "name": "default",
    "window": {
        "designWidth": 720,
        "autoDesignWidth": false
```

            }
        }
    ]

(2) 新建 climateAreaList 组件作为页面入口组件，climateListItem 为其子组件。

```
@Component
struct climateListItem {
    build() {}
}

@Entry
@Component
struct climateAreaList {
    build() {
        List() {
            ListItem() {
                climateListItem()
            }
        }
    }
}
```

(3) 引入 climateData 类和 initializeOnStartup 方法。

```
import { climateData } from '../model/climateData'
import { initializeOnStartup } from '../model/climateDataModels'
```

(4) climateAreaList 和 climateListItem 组件数值传递。在 climateAreaList 组件内创建类型为 climateData[ ]的成员变量 climateItems，调用 initializeOnStartup()方法为其赋值。在 climateListItem 组件内创建类型为 climateData 的成员变量 climateItem。将父组件 climateItems 数组的第一个子组件的 climateItems[0]作为参数传递给 climateListItem。

(5) 声明子组件 climateListItem 的 UI 布局。

(6) 创建两个 climateListItem。在 List 组件下创建两个 climateListItem，分别给 climateListItem 传递 climateItems 数组的第一个子组件 this.climateItems[0]和第二个子组件 this.climateItems[1]。

(7) 单独创建每一个 climateListItem 肯定是不合理的。这就需要引入 ForEach 循环渲染。

```
\\climateAreaList.ets
import { climateData } from '../model/climateData'
import { initializeOnStartup } from '../model/climateDataModels'
import { Category } from '../model/climateData'


@Component
struct climateListItem {
    private climateItem: climateData
    build() {
```

```
        Flex({ justifyContent: FlexAlign.Start, alignItems: ItemAlign.Center }) {
            Image(this.climateItem.image)
                .objectFit(ImageFit.Contain)
                .height(40)
                .width(40)
                .backgroundColor('#FFf1f3f5')
                .margin({ right: 16 })
            Text(this.climateItem.name)
                .fontSize(14)
                .flexGrow(1)
            Text(this.climateItem.category + '线城市')
                .fontSize(14)
        }
        .height(64)
        .margin({ right: 24, left: 32 })
    }
}

@Entry
@Component
struct climateAreaList {
    private climateItems: climateData[] = initializeOnStartup()
    build() {
        Column() {
            Flex({justifyContent: FlexAlign.Start, alignItems: ItemAlign.Center}) {
                Text('城市列表')
                    .fontSize(20)
                    .margin({ left: 20 })
            }
            .height('7%')
            .backgroundColor('#FFf1f3f5')
            List() {
                //ListItem() {
                //    climateListItem({ climateItem: this.climateItems[0] })
                //}
                //ListItem() {
                //    climateListItem({ climateItem: this.climateItems[1] })
                //}
                ForEach(this.climateItems, item => {
                    ListItem() {
                        climateListItem({ climateItem: item })
                    }
                }, item => item.id.toString())
            }
            .height('93%')
        }
    }
}
```

4. Grid 布局

应用在主页提供列表 List 和网格 Grid 两种城市显示方式。网格布局方式如图 4.25 所示。

图 4.25　网格布局方式

（1）将 Category 枚举类型引入 climateAreaList 页面。

import { Category} from '../model/climateData'

（2）创建 climateAreaCategoryList 和 climateAreaCategory 组件，其中，climateAreaCategoryList 作为新的页面入口组件，在入口组件调用 initializeOnStartup（）方法。在 climateAreaCategoryList 组件内创建 showList 成员变量，用于控制 List 布局和 Grid 布局的渲染切换。Stack 为堆叠组件，需要用到条件渲染语句 if…else…。

（3）在页面右上角创建切换 List/Grid 布局的图标。设置 Stack 对齐方式为顶部尾部对齐 TopEnd，创建 Image 组件，设置其单击事件，即 showList 取反。添加 @State 装饰器。单击右上角的 switch 标签后，页面没有任何变化，这是因为 showList 不是有状态数据，它的改变不会触发页面的刷新。需要为其添加 @State 装饰器，使其成为状态数据，它的改变会引起其所在组件的重新渲染。

```
@Component
struct climateAreaCategory {
    private climateAreaItems: climateData[]
    build() {

    }
}
```

```
@Entry
@Component
struct climateAreaCategoryList {
        private climateAreaItems: climateData[] = initializeOnStartup()
        @State private showList: boolean = false /@State 修饰器
        build() {
                Stack({ alignContent: Alignment.TopEnd }) {
                        if (this.showList) {
                                climateAreaList({ climateItems: this.climateAreaItems })
                        } else {
                                climateAreaCategory({ climateAreaItems: this.climateAreaItems })
                        }
                        Image( $ r('app.media.Switch'))
                        .height(24)
                        .width(24)
                        .margin({ top: 15, right: 10 })
                        .onClick(() => {
                                this.showList = !this.showList
                        })
                }.height('100 % ')
        }
}
```

（4）创建显示所有食物的页签（All）。在 climateAreaCategory 组件内创建 Tabs 组件和其子组件 TabContent，设置 tabBar 为 All。设置 TabBars 的宽度为 280，布局模式为 Scrollable，即超过总长度后可以滑动。Tabs 是一种可以通过页签进行内容视图切换的容器组件，每个页签对应一个内容视图 TabContent。

（5）创建 climateGrid 组件，作为 TabContent 的子组件。

（6）实现 2×3 的网格布局。创建 Grid 组件，设置列数为 columnsTemplate('1fr 1fr')，行数为 rowsTemplate('1fr 1fr 1fr')，行间距 rowsGap 和列间距 columnsGap 为 8。创建 Scroll 组件，使其可以滑动。

（7）创建 climateGridItem 组件，展示城市图片、名称和类别，实现其 UI 布局，为 GridItem 的子组件。每个 climateGridItem 高度为 184，行间距为 8，设置 Grid 总高度为 (184＋8)×6－8＝1144。

（8）创建展示城市分类的页签。

（9）设置不同城市分类的 Grid 的行数和高度。因为不同类别城市数量不同，所以不能用'1fr 1fr 1fr'常量来统一设置成 3 行。创建 gridRowTemplate 和 HeightValue 成员变量，通过成员变量设置 Grid 行数和高度。调用 aboutToAppear 接口计算行数（gridRowTemplate）和高度（heightValue）。

```
//climateAreaList.ets
import { climateData } from '../model/climateData'
import { initializeOnStartup } from '../model/climateDataModels'
import { Category } from '../model/climateData'
import router from '@system.router'
```

```
@Component
struct climateListItem {
        private climateItem: climateData
        build() {
                Navigator({ target: 'pages/climateDetail' }) {
                        Flex ({ justifyContent: FlexAlign. Start, alignItems: ItemAlign. Center }) {
                                Image(this.climateItem.image)
                                        .objectFit(ImageFit.Contain)
                                        .height(40)
                                        .width(40)
                                        .backgroundColor('#FFf1f3f5')
                                        .margin({ right: 16 })
                                Text(this.climateItem.name)
                                        .fontSize(14)
                                        .flexGrow(1)
                                Text(this.climateItem.category + '线城市')
                                        .fontSize(14)
                        }
                        .height(64)
                        .margin({ right: 24, left: 32 })
                }.params({ climateData: this.climateItem })
        }
}


@Component
struct climateAreaList {
        private climateItems: climateData[] = initializeOnStartup()
        build() {
                Column() {
                        Flex ({ justifyContent: FlexAlign. Start, alignItems: ItemAlign. Center}) {
                                Text('城市列表')
                                        .fontSize(20)
                                        .margin({ left: 20 })
                        }
                        .height('7%')
                        .backgroundColor('#FFf1f3f5')
                        List() {
                                //ListItem() {
                                        //climateListItem({ climateItem: this.climateItems[0] })
                                //}
                                //ListItem() {
                                        //climateListItem({ climateItem: this.climateItems[1] })
                                //}
                                ForEach(this.climateItems, item => {
                                        ListItem() {
                                                climateListItem({ climateItem: item })
```

```
                    }
                }, item => item.id.toString())
            }
            .height('93%')
        }
    }
}
@Component
struct climateGridItem {
    private climateItem: climateData
    build() {
        Column() {
            Row() {
                Image(this.climateItem.image)
                    .objectFit(ImageFit.Contain)
                    .height(152)
                    .width('100%')
            }.backgroundColor('#FFf1f3f5')
            Flex({ justifyContent: FlexAlign.Start, alignItems: ItemAlign.Center }) {
                Text(this.climateItem.name)
                    .fontSize(14)
                    .flexGrow(1)
                    .padding({ left: 8 })
                Text(this.climateItem.category + '线城市')
                    .fontSize(14)
                    .margin({ right: 6 })
            }
            .height(32)
            .width('100%')
            .backgroundColor('#FFe5e5e5')
        }
        .height(184)
        .width('100%')
        .onClick(() => {
            router.push({ uri: 'pages/climateDetail',
                params: { climateData: this.climateItem }})
        })
    }
}

@Component
struct climateGrid {
    private climateItems: climateData[]
    private gridRowTemplate : string = ''
    private heightValue: number
    aboutToAppear() {
        var rows = Math.round(this.climateItems.length / 2);
        this.gridRowTemplate = '1fr '.repeat(rows);
        this.heightValue = rows * 192 - 8;
    }
```

```
build() {
    Scroll() {
        Grid() {
            ForEach(this.climateItems, (item: climateData) => {
                GridItem() {
                    climateGridItem({climateItem: item})
                }
            }, (item: climateData) => item.id.toString())
        }
        //.rowsTemplate('1fr 1fr 1fr')
        .rowsTemplate(this.gridRowTemplate)
        .columnsTemplate('1fr 1fr')
        .columnsGap(8)
        .rowsGap(8)
        //.height(1144)
        .height(this.heightValue)
    }
    .scrollBar(BarState.Off)
    .padding({left: 16, right: 16})
}
}

@Component
struct climateAreaCategory {
    private climateAreaItems: climateData[]
    build() {
        Stack() {
            Tabs() {
                TabContent() {
                    climateGrid({ climateItems: this.climateAreaItems })
                }
                .tabBar('All')
                TabContent() {
                    climateGrid({ climateItems: this.climateAreaItems.filter(item => (item.category === Category.cityclass0)) })
                }.tabBar('0 线城市')

                TabContent() {
                    climateGrid({ climateItems: this.climateAreaItems.filter(item => (item.category === Category.cityclass1)) })
                }.tabBar('1 线城市')

                TabContent() {
                    climateGrid({ climateItems: this.climateAreaItems.filter(item => (item.category === Category.cityclass2)) })
                }.tabBar('2 线城市')

                TabContent() {
                    climateGrid({ climateItems: this.climateAreaItems.filter(item => (item.category === Category.cityclass3)) })
                }.tabBar('3 线城市')
```

```
                                        TabContent() {
                                                climateGrid({ climateItems: this.climateAreaItems.
filter(item => (item.category === Category.cityclass4)) })
                                        }.tabBar('4 线城市')

                                        TabContent() {
                                                climateGrid({ climateItems: this.climateAreaItems.
filter(item => (item.category === Category.cityclass5)) })
                                        }.tabBar('5 线城市')
                                }
                                .barWidth(280)
                                .barMode(BarMode.Scrollable)
                        }
                }
        }
}

@Entry
@Component
struct climateAreaCategoryList {
        private climateAreaItems: climateData[] = initializeOnStartup()
        @State private showList: boolean = false
        build() {
                Stack({ alignContent: Alignment.TopEnd }) {
                        if (this.showList) {
                                climateAreaList({ climateItems: this.climateAreaItems })
                        } else {
                                climateAreaCategory({ climateAreaItems: this.climateAreaItems })
                        }
                        Image( $ r('app.media.Switch'))
                        .height(24)
                        .width(24)
                        .margin({ top: 15, right: 10 })
                        .onClick(() => {
                                this.showList = !this.showList
                        })
                }.height('100 %')
        }
}
```

5．页面跳转

声明式 UI 范式提供了路由容器组件 Navigator 和路由 RouterAPI 接口两种机制来实现页面间的跳转。前者包装了页面路由的能力，指定页面 target 后，使其包裹的子组件都具有路由能力。后者通过在页面上引入 router，可以调用 router 的各种接口，从而实现页面路由的各种操作。

（1）列表布局使用 Navigator 组件。在 climateListItem 内创建 Navigator 组件，使其子组件都具有路由功能，目标页面 target 为'pages/climateDetail'。

（2）Grid 布局使用 Router 接口。单击 climateGridItem 后跳转到 climateDetail 页面。

调用页面路由 router 模块的 push 接口,将 climateDetail 页面推到路由栈中,实现页面跳转。使用 router 路由 API,需要先引入 router。

(3) 在 climateDetail 页面增加回到城市列表页面的图标。在 resources→phone→media 目录下存入回退图标 Back.png。新建自定义组件 PageTitle,包含后退的图标和 climate Detail 的文本,调用路由的 router.back()接口,弹出路由栈最上面的页面,即返回上一级页面。

(4) 接下来构建两个页面的数据传递,需要用到携带参数(parameter)路由。在 climateListItem 组件的 Navigator 设置其 params 属性,params 属性接受 key-value 的 Object。climateGridItem 调用的 routerAPI 同样有携带参数跳转的能力,使用方法和 Navigator 类似。

(5) climateDetail 页面引入 climateData 类,在 climateDetail 组件内添加 climateItem 成员变量。

(6) 获取 climateData 对应的 value。调用 router.getParams().climateData 获取到 climateAreaList 页面跳转来时携带的 climateDate 对应的数据。

(7) 重构 climateDetail 页面的组件。在构建视图时,climateDetail 页面的信息都是直接声明的常量,现在要用传递来的 climateData 数据来对其进行重新赋值。整体的 climateDetail.ets 代码如下。

```
//climateDetail.ets
import router from '@system.router'
import { climateData } from '../model/climateData'

@Component
struct PageTitle {
    build() {
        Flex({ alignItems: ItemAlign.Start }) {
            Image( $r('app.media.Back'))
                .width(21.8)
                .height(19.6)
            Text('气候特性')
                .fontSize(21.8)
                .margin({left: 17.4})
        }
        .height(61)
        .backgroundColor('#FFedf2f5')
        .padding({ top: 13, bottom: 15, left: 28.3 })
        .onClick(() => {
            router.back()
        })
    }
}
@Component
struct ImageDisplay {
    private climateItem: climateData
    build() {
        Stack({ alignContent: Alignment.BottomStart }) {
```

```
                    Image(this.climateItem.image)
                        .objectFit(ImageFit.Contain)
                    Text(this.climateItem.name)
                        .fontSize(26)
                        .fontWeight(500)
                        .margin({ left: 26, bottom: 17.4 })
            }
            .height(357)
            .backgroundColor('#FFedf2f5')
    }
}

@Component
struct ContentTable {
    private climateItem: climateData

    @Builder climateParaItem(title: string, name: string, value: string) {
        Flex() {
            Text(title)
                .fontSize(17.4)
                .fontWeight(FontWeight.Bold)
                .layoutWeight(1)
            Flex() {
                Text(name)
                    .fontSize(17.4)
                    .flexGrow(1)
                Text(value)
                    .fontSize(17.4)
            }
            .layoutWeight(2)
        }
    }

    build() {
        Flex ({ direction: FlexDirection.Column, justifyContent: FlexAlign.SpaceBetween, alignItems: ItemAlign.Start }) {
                this.climateParaItem('', '坐标', '东经' + this.climateItem.latitude + '度')
                this.climateParaItem('', '', '北纬' + this.climateItem.longitude + '度')
                this.climateParaItem('', '最高温度', this.climateItem.temceiling + '摄氏度')
                this.climateParaItem('', '最低温度', this.climateItem.tembottom + '摄氏度')
                this.climateParaItem('', '降水量', this.climateItem.precipitation + 'ml')
                this.climateParaItem('', '空气质量', this.climateItem.airquality + '')
        }
        .height(280)
        .padding({ top: 30, right: 30, left: 30 })
    }
```

```
    }

    @Entry
    @Component
    struct climateDetail {
        private climateItem: climateData = router.getParams().climateData
        build() {
            Column() {
                Stack( { alignContent: Alignment.TopStart }) {
                    ImageDisplay({ climateItem: this.climateItem })
                    PageTitle()
                }
                ContentTable({ climateItem: this.climateItem })
            }
            .alignItems(HorizontalAlign.Center)
        }
    }
```

4.22 WebSocket 客户端实验

1. 实验内容

本实验创建和 WebSocket 服务器连接的应用程序。练习 WebSocket 的创建、连接和事件等接口的使用。

2. 实验代码

本实验可以通过 JavaScript 开发范式或者 eTS 开发范式进行。这里新建一个低代码开发方式的工程,在该工程的 index.js 中,已经放置了 Text 组件,Text 组件有一 click 事件,该 click 事件的功能是改变 Text 的 Value 属性,显示不同的文字。把 switchTitle()函数做一修改,代码如下。

```
\\index.js
export default {
    data: {

        title: "Hello HarmonyOS",
        isHarmonyOS: true,
    },


    switchTitle() {
        let that = this;
        that.title = that.isHarmonyOS ? "Hello World" : "Hello HarmonyOS";
        that.isHarmonyOS = !that.isHarmonyOS;

        let ws = webSocket.createWebSocket();

        let url1 = "ws://121.40.165.18: 8800";
```

```
            ws.on('open', (err, value) => {
                console.log("on open, status: " + value.status + ", message: " +
value.message);
                //当收到 on('open')事件时,可以通过 send()方法与服务器进行通信
                ws.send("Hello, server!", (err, value) => {
                    if (!err) {
                        console.log("send success");
                    } else {
                        console.log("send fail, err: " + JSON.stringify
(err));
                    }
                });
            });
            ws.on('message', (err, value) => {
                console.log("on message, message: " + value);
                this.title = value;
                /* 当收到服务器的bye消息时(此消息字段仅为示意,具体字段需要与
服务器协商),主动断开连接 */
                if (value === 'bye') {
                    ws.close();
                    Promise.then((value) => {
                        console.log("close success");
                    }).catch((err) => {
                        console.log("close fail, err is " + JSON.
stringify(err));
                    });
                }
            });
            ws.on('close', (err, value) => {
                console.log("websocket 关闭, code is " + value.code + ", reason is " +
value.reason);
            });
            ws.on('error', (err) => {
                console.log("websocket 错误, error: " + JSON.stringify(err));
            });

            let promise = ws.connect(url);
            promise.then((value) => {
                console.log("connect success")
            }).catch((err) => {
                console.log("connect fail, error: " + JSON.stringify(err))
            });
        }
}
```

使用 WebSocket 之前需要先 import,还要注意需要申请访问权限。本例的 config.json 中,权限设置如下,其中第二个对于本实验是必需的。

```
"reqPermissions": [
{
```

```
                "name": "ohos.permission.GET_NETWORK_INFO"
        },
        {
                "name": "ohos.permission.INTERNET"
        },
        {
                "name": "ohos.permission.SET_NETWORK_INFO"
        },
        {
                "name": "ohos.permission.MANAGE_WIFI_CONNECTION"
        },
        {
                "name": "ohos.permission.SET_WIFI_INFO"
        },
        {
                "name": "ohos.permission.GET_WIFI_INFO"
        }
]
```

3. 实验结果

启动手机模拟器,启动成功后运行应用。运行后单击 Text 组件即可创建连接并接收来自 WebSocket 服务器的消息。

4.23 MQTT 客户端实验

1. 实验内容

在本实验中,通过 WebSocket 实现 MQTT 包的发送接收,其中包括两部分实验。一个是自行构建 MQTT 协议的包实现客户端,另一个是通过开源 JavaScript 的 MQTT 客户端库实现客户端。

2. 实验代码

开源的 MQTT 协议客户端库提供了 MQTT 客户端各种操作的接口。paho-mqtt 是其中一个,其源码可以在 https://github.com/eclipse/paho.mqtt.javascript.git 下载。于是 MQTT 的操作均可以通过调用该接口实现。如下为一段 HTML 和 JavaScript 代码,需要浏览器打开。

```html
<!-- index.html -->
<!DOCTYPE html>
<html>
<head>
<meta charset="utf-8">
<title>MQTT</title>
<script>
function displayDate(){
        document.getElementById("demo").innerHTML = Date();
}
```

```html
</script>
<!-- mqttws31.js 为 paho mqtt 客户端库文件 -->
<script src="C:\Users\fg\AppData\Roaming\npm\node_modules\pahomqtt\mqttws31.js"></script>
</head>
<body>

<script src="index.js"></script>

<h1>MQTT 接收消息</h1>
<p id="demo">显示日期</p>
<p id="demo2">显示接收到的消息</p>

<button type="button" onclick="displayDate()">显示日期</button>
<button type="button" onclick="pub()">发送接收</button>

</body>
</html>
```

```javascript
//index.js
{
    //创建实例
    client = new Paho.Client('broker.emqx.io', 8083, "clientId");

    //设置回调函数
    client.onConnectionLost = onConnectionLost;
    client.onMessageArrived = onMessageArrived;

    //连接客户端
    client.connect({onSuccess: onConnect});


    //连接时调用
    function onConnect() {
        //建立连接后,订阅一个主题 test,并发送一条消息
        console.log("onConnect");
        client.subscribe("test");                //订阅主题
        message = new Paho.Message("Hello 你好こんにちはBonjour!");
        message.destinationName = "test";
        client.send(message);                    //发送消息
    }

    //关闭连接时调用
    function onConnectionLost(responseObject) {
        if (responseObject.errorCode !== 0) {
            console.log("onConnectionLost: " + responseObject.errorMessage);
        }
    }

    //收到消息时调用
    function onMessageArrived(message) {
```

```
            console.log("onMessageArrived: " + message.payloadString);
            document.getElementById("demo2").innerHTML = message.payloadString;
        }
    }
```

3. 实验结果

该客户端和服务器的交互过程如图 4.26 所示。其中最后两条消息是两个字节的 ping 包和 pong 包。

图 4.26　paho mqtt 客户端库示例

开源 JavaScript 的 MQTT 客户端库通过浏览器的 WebSocket 实现,前述已知 HarmonyOS 提供的 WebSocket 模块和浏览器的 WebSocket 不同。如果在 HarmonyOS 中使用 paho mqtt 客户端库,需要对其中的 WebSocket 的使用进行一些修改。

附录A

WiFi IoT核心板GPIO配置

WiFi IoT 主板为 Hi3861 核心板。Hi3861 主控芯片是一款高度集成的 2.4GHz WLAN SoC 芯片，集成 IEEE 802.11b/g/n 基带和 RF 电路。Hi3861 最大工作频率为 160MHz，外设接口包括 2 个 SPI 接口、2 个 I2C 接口、3 个 UART 接口、15 个 GPIO 接口、7 路 ADC 输入、6 路 PWM、1 个 I2S 接口，同时支持高速 SDIO 2.0 接口，最高时钟可达 50MHz。

Hi3861 的 GPIO 端口有 15 个，名称分别为 GPIO_00～GPIO_14，所有 IO 都可作为输入输出口，电平为 3.3V 或 1.8V，都有防倒灌功能。其中，GPIO_03、GPIO_05、GPIO_07 和 GPIO_14 在 Udsleep(ultra deep sleep)模式下，这些 IO 端口上升沿可触发唤醒，其 GPIO 符合 AMBA2.0 的 APB 协议。主要接口有 APB 接口、IO pad 的外部数据接口和中断信号接口。

Hi3861 的 GPIO 端口使用情况如表 A.1 所示。

表 A.1 Hi3861 的 GPIO 端口配置

	核心板	显示板	NFC板	环境监测板	智能红绿灯板	智能(炫彩)灯板	机器人板
GPIO00	—	—	—	—	—	—	电机驱动芯片1的A路PWM信号（PWM3_OUT）
GPIO01	—	—	—	—	—	—	电机驱动芯片1的B路PWM信号（PWM4_OUT）
GPIO02	—	—	—	—	—	—	舵机接口PWM信号（PWM2_OUT）
GPIO03	串口TX信号(UART0_TX)	—	—	—	—	—	—

续表

	核心板	显示板	NFC 板	环境监测板	智能红绿灯板	智能(炫彩)灯板	机器人板
GPIO04	串口 RX 信号(UART0_RX)	—	—	—	—	—	—
GPIO05	按键 S2 状态信号(ADC2)	按键 S1 和按键 S2 状态信号(SWITCH)(ADC2)					串口接 RX 信号（UART1_RXD）
GPIO06	—	—	—	—	—	—	串口接 TX 信号（UART1_TXD）
GPIO07	—	—	—	—	人体红外感应器检知信号（REL）(GPIO07 或 ADC3)	—	超声波接口触发信号（Trig）(GPIO07 或 PWM0_OUT)
GPIO08	—	—	—	—	按键 S1 状态信号（SWITCH）(GPIO08)	—	超声波接口回信号（Echo）(GPIO08)
GPIO09	LED1 开关(GPIO09)	MOSI 引脚(SPI0_TXD)	NFC 芯片 CSN 信号(GPIO09)	蜂鸣器控制信号(BEEP)(GPIO09 或 PWM0_OUT)	蜂鸣器控制信号(BEEP)(GPIO09 或 PWM0_OUT)	光敏电阻检知信号(PHO_RES)(ADC4)	电动机驱动芯片 2 的 A 路 PWM 信号(PWM0_OUT)
GPIO10	—	CLK 引脚(SPI0_CLK)	IRQ 引脚(GPIO10)	—	红灯控制引脚（RED）(GPIO10 或 PWM1_OUT)	红灯控制引脚（RED）(GPIO10 或 PWM1_OUT)	电动机驱动芯片 2 的 A 路 PWM 信号(PWM0_OUT)
GPIO11	—	MISO 引脚(SPI0_RXD)	—	可燃气体传感器检测结果信号(ADC5)	绿灯控制引脚(GREEN)(GPIO11 或 PWM2_OUT)	绿灯控制引脚(GREEN)(GPIO11 或 PWM2_OUT)	传感器 1 接口输入信号（GPIO11 或 ADC5）

续表

	核心板	显示板	NFC板	环境监测板	智能红绿灯板	智能(炫彩)灯板	机器人板
GPIO12	—	GPIO12	—	—	黄灯控制引脚(YELLOW)(GPIO12 或 PWM3_OUT)	蓝灯控制引脚(BLU7/E)(GPIO12 或 PWM3_OUT)	传感器2接口输入信号(GPIO12 或 ADC0)
GPIO13	—	I2C 数据线(I2C0_SDA)	I2C 数据线(I2C0_SDA)	温湿度传感器 I2C 数据线(I2C0_SDA)	—	—	I2C 接口1和I2C 接口2数据线(I2C0_SDA)
GPIO14	—	I2C 时钟线(I2C0_SCL)	I2C 时钟线(I2C0_SCL)	温湿度传感器 I2C 时钟线(I2C0_SCL)	—	—	I2C 接口1和I2C 接口2时钟线(I2C0_SCL)

附录B

GPIO扩展功能源代码文件

OpenHarmony 3.0 LTS 中，wifiiot_gpio_ex.h 缺失，如果使用该文件，需要自行添加到相应目录中。同时需要添加相应的 C 文件。这些文件放在目录 base/iot_hardware/peripheral/interfaces/kits 下，该目录下的 BUILD.gn 文件也需要修改。

这三个文件的源代码如下。

B.1 wifiiot_gpio_ex.h

```
/*
 * @file iot_gpio_ex.h
 *
 * @brief 声明扩展 GPIO 接口函数
 *
 * 这些函数用于设置 GPIO 拉取和驱动能力
 *
 * @since 1.0
 * @version 1.0
 */

#ifndef IOT_GPIO_EX_H
#define IOT_GPIO_EX_H

/*
 * @brief GPIO 上拉或下拉设置的枚举
 */
typedef enum {
    /** 无设置 */
    IOT_IO_PULL_NONE,
    /** 上拉 */
    IOT_IO_PULL_UP,
    /** 下拉 */
    IOT_IO_PULL_DOWN,
    /** 最大值 */
    IOT_IO_PULL_MAX,
```

```c
} IotIoPull;

/*
 * @ingroup iot_io
 *
 * IO 硬件管脚编号
 */
typedef enum {
    IOT_IO_NAME_GPIO_0,      /**< GPIO0 */
    IOT_IO_NAME_GPIO_1,      /**< GPIO1 */
    IOT_IO_NAME_GPIO_2,      /**< GPIO2 */
    IOT_IO_NAME_GPIO_3,      /**< GPIO3 */
    IOT_IO_NAME_GPIO_4,      /**< GPIO4 */
    IOT_IO_NAME_GPIO_5,      /**< GPIO5 */
    IOT_IO_NAME_GPIO_6,      /**< GPIO6 */
    IOT_IO_NAME_GPIO_7,      /**< GPIO7 */
    IOT_IO_NAME_GPIO_8,      /**< GPIO8 */
    IOT_IO_NAME_GPIO_9,      /**< GPIO9 */
    IOT_IO_NAME_GPIO_10,     /**< GPIO10 */
    IOT_IO_NAME_GPIO_11,     /**< GPIO11 */
    IOT_IO_NAME_GPIO_12,     /**< GPIO12 */
    IOT_IO_NAME_GPIO_13,     /**< GPIO13 */
    IOT_IO_NAME_GPIO_14,     /**< GPIO14 */
    IOT_IO_NAME_SFC_CSN,     /**< SFC_CSN */
    IOT_IO_NAME_SFC_IO1,     /**< SFC_IO1 */
    IOT_IO_NAME_SFC_IO2,     /**< SFC_IO2 */
    IOT_IO_NAME_SFC_IO0,     /**< SFC_IO0 */
    IOT_IO_NAME_SFC_CLK,     /**< SFC_CLK */
    IOT_IO_NAME_SFC_IO3,     /**< SFC_IO3 */
    IOT_IO_NAME_MAX,
} IotIoName;

/*
 * @ingroup iot_io
 *
 * GPIO_0 管脚功能
 */
typedef enum {
    IOT_IO_FUNC_GPIO_0_GPIO,
    IOT_IO_FUNC_GPIO_0_UART1_TXD = 2,
    IOT_IO_FUNC_GPIO_0_SPI1_CK,
    IOT_IO_FUNC_GPIO_0_JTAG_TDO,
    IOT_IO_FUNC_GPIO_0_PWM3_OUT,
    IOT_IO_FUNC_GPIO_0_I2C1_SDA,
} IotIoFuncGpio0;

/*
 * @ingroup iot_io
 *
 * GPIO_1 管脚功能
 */
```

```c
typedef enum {
        IOT_IO_FUNC_GPIO_1_GPIO,
        IOT_IO_FUNC_GPIO_1_UART1_RXD = 2,
        IOT_IO_FUNC_GPIO_1_SPI1_RXD,
        IOT_IO_FUNC_GPIO_1_JTAG_TCK,
        IOT_IO_FUNC_GPIO_1_PWM4_OUT,
        IOT_IO_FUNC_GPIO_1_I2C1_SCL,
        IOT_IO_FUNC_GPIO_1_BT_FREQ,
} IotIoFuncGpio1;

/*
 * @ingroup iot_io
 *
 * GPIO_2 管脚功能
 */
typedef enum {
        IOT_IO_FUNC_GPIO_2_GPIO,
        IOT_IO_FUNC_GPIO_2_UART1_RTS_N = 2,
        IOT_IO_FUNC_GPIO_2_SPI1_TXD,
        IOT_IO_FUNC_GPIO_2_JTAG_TRSTN,
        IOT_IO_FUNC_GPIO_2_PWM2_OUT,
        IOT_IO_FUNC_GPIO_2_SSI_CLK = 7,
} IotIoFuncGpio2;

/*
 * @ingroup iot_io
 *
 * GPIO_3 管脚功能
 */
typedef enum {
        IOT_IO_FUNC_GPIO_3_GPIO,
        IOT_IO_FUNC_GPIO_3_UART0_TXD,
        IOT_IO_FUNC_GPIO_3_UART1_CTS_N,
        IOT_IO_FUNC_GPIO_3_SPI1_CSN,
        IOT_IO_FUNC_GPIO_3_JTAG_TDI,
        IOT_IO_FUNC_GPIO_3_PWM5_OUT,
        IOT_IO_FUNC_GPIO_3_I2C1_SDA,
        IOT_IO_FUNC_GPIO_3_SSI_DATA,
} IotIoFuncGpio3;

/*
 * @ingroup iot_io
 *
 * GPIO_4 管脚功能
 */
typedef enum {
        IOT_IO_FUNC_GPIO_4_GPIO,
        IOT_IO_FUNC_GPIO_4_UART0_RXD = 2,
        IOT_IO_FUNC_GPIO_4_JTAG_TMS = 4,
        IOT_IO_FUNC_GPIO_4_PWM1_OUT,
        IOT_IO_FUNC_GPIO_4_I2C1_SCL,
```

```c
} IotIoFuncGpio4;

/*
 * @ingroup iot_io
 *
 * GPIO_5 管脚功能
 */
typedef enum {
        IOT_IO_FUNC_GPIO_5_GPIO,
        IOT_IO_FUNC_GPIO_5_UART1_RXD = 2,
        IOT_IO_FUNC_GPIO_5_SPI0_CSN,
        IOT_IO_FUNC_GPIO_5_PWM2_OUT = 5,
        IOT_IO_FUNC_GPIO_5_I2S0_MCLK,
        IOT_IO_FUNC_GPIO_5_BT_STATUS,
} IotIoFuncGpio5;

/*
 * @ingroup iot_io
 *
 * GPIO_6 管脚功能
 */
typedef enum {
        IOT_IO_FUNC_GPIO_6_GPIO,
        IOT_IO_FUNC_GPIO_6_UART1_TXD = 2,
        IOT_IO_FUNC_GPIO_6_SPI0_CK,
        IOT_IO_FUNC_GPIO_6_PWM3_OUT = 5,
        IOT_IO_FUNC_GPIO_6_I2S0_TX,
        IOT_IO_FUNC_GPIO_6_COEX_SWITCH,
} IotIoFuncGpio6;

/*
 * @(ingroup) iot_io
 *
 * GPIO_7 管脚功能
 */
typedef enum {
        IOT_IO_FUNC_GPIO_7_GPIO,
        IOT_IO_FUNC_GPIO_7_UART1_CTS_N = 2,
        IOT_IO_FUNC_GPIO_7_SPI0_RXD,
        IOT_IO_FUNC_GPIO_7_PWM0_OUT = 5,
        IOT_IO_FUNC_GPIO_7_I2S0_BCLK,
        IOT_IO_FUNC_GPIO_7_BT_ACTIVE,
} IotIoFuncGpio7;

/*
 * @ingroup iot_io
 *
 * GPIO_8 管脚功能
 */
typedef enum {
        IOT_IO_FUNC_GPIO_8_GPIO,
```

```c
        IOT_IO_FUNC_GPIO_8_UART1_RTS_N = 2,
        IOT_IO_FUNC_GPIO_8_SPI0_TXD,
        IOT_IO_FUNC_GPIO_8_PWM1_OUT = 5,
        IOT_IO_FUNC_GPIO_8_I2S0_WS,
        IOT_IO_FUNC_GPIO_8_WLAN_ACTIVE,
} IotIoFuncGpio8;

/*
 * @ingroup iot_io
 *
 * GPIO_9 管脚功能
 */
typedef enum {
        IOT_IO_FUNC_GPIO_9_GPIO,
        IOT_IO_FUNC_GPIO_9_I2C0_SCL,
        IOT_IO_FUNC_GPIO_9_UART2_RTS_N,
        IOT_IO_FUNC_GPIO_9_SDIO_D2,
        IOT_IO_FUNC_GPIO_9_SPI0_TXD,
        IOT_IO_FUNC_GPIO_9_PWM0_OUT,
        IOT_IO_FUNC_GPIO_9_I2S0_MCLK = 7,
} IotIoFuncGpio9;

/*
 * @ingroup iot_io
 *
 * GPIO_10 管脚功能
 */
typedef enum {
        IOT_IO_FUNC_GPIO_10_GPIO,
        IOT_IO_FUNC_GPIO_10_I2C0_SDA,
        IOT_IO_FUNC_GPIO_10_UART2_CTS_N,
        IOT_IO_FUNC_GPIO_10_SDIO_D3,
        IOT_IO_FUNC_GPIO_10_SPI0_CK,
        IOT_IO_FUNC_GPIO_10_PWM1_OUT,
        IOT_IO_FUNC_GPIO_10_I2S0_TX = 7,
} IotIoFuncGpio10;

/*
 * @ingroup iot_io
 *
 * GPIO_11 管脚功能
 */
typedef enum {
        IOT_IO_FUNC_GPIO_11_GPIO,
        IOT_IO_FUNC_GPIO_11_UART2_TXD = 2,
        IOT_IO_FUNC_GPIO_11_SDIO_CMD,
        IOT_IO_FUNC_GPIO_11_SPI0_RXD,
        IOT_IO_FUNC_GPIO_11_PWM2_OUT,
        IOT_IO_FUNC_GPIO_11_RF_TX_EN_EXT,
        IOT_IO_FUNC_GPIO_11_I2S0_RX,
} IotIoFuncGpio11;
```

```c
/*
 * @ingroup iot_io
 *
 * GPIO_12 管脚功能
 */
typedef enum {
        IOT_IO_FUNC_GPIO_12_GPIO,
        IOT_IO_FUNC_GPIO_12_UART2_RXD = 2,
        IOT_IO_FUNC_GPIO_12_SDIO_CLK,
        IOT_IO_FUNC_GPIO_12_SPI0_CSN,
        IOT_IO_FUNC_GPIO_12_PWM3_OUT,
        IOT_IO_FUNC_GPIO_12_RF_RX_EN_EXT,
        IOT_IO_FUNC_GPIO_12_I2S0_BCLK,
} IotIoFuncGpio12;

/*
 * @ingroup iot_io
 *
 * GPIO_13 管脚功能
 */
typedef enum {
        IOT_IO_FUNC_GPIO_13_SSI_DATA,
        IOT_IO_FUNC_GPIO_13_UART0_TXD,
        IOT_IO_FUNC_GPIO_13_UART2_RTS_N,
        IOT_IO_FUNC_GPIO_13_SDIO_D0,
        IOT_IO_FUNC_GPIO_13_GPIO,
        IOT_IO_FUNC_GPIO_13_PWM4_OUT,
        IOT_IO_FUNC_GPIO_13_I2C0_SDA,
        IOT_IO_FUNC_GPIO_13_I2S0_WS,
} IotIoFuncGpio13;

/*
 * @ingroup iot_io
 *
 * GPIO_14 管脚功能
 */
typedef enum {
        IOT_IO_FUNC_GPIO_14_SSI_CLK,
        IOT_IO_FUNC_GPIO_14_UART0_RXD,
        IOT_IO_FUNC_GPIO_14_UART2_CTS_N,
        IOT_IO_FUNC_GPIO_14_SDIO_D1,
        IOT_IO_FUNC_GPIO_14_GPIO,
        IOT_IO_FUNC_GPIO_14_PWM5_OUT,
        IOT_IO_FUNC_GPIO_14_I2C0_SCL,
} IotIoFuncGpio14;

/*
 * @ingroup iot_io
 *
 * SFC_CSN 管脚功能
```

```c
 */
typedef enum {
    IOT_IO_FUNC_SFC_CSN_SFC_CSN,
    IOT_IO_FUNC_SFC_CSN_SDIO_D2,
    IOT_IO_FUNC_SFC_CSN_GPIO9,
    IOT_IO_FUNC_SFC_CSN_SPI0_TXD = 4,
} IotIoFuncSfcCsn;

/*
 * @ingroup iot_io
 *
 * SFC_DO 管脚功能
 */
typedef enum {
    IOT_IO_FUNC_SFC_IO_1_SFC_DO,
    IOT_IO_FUNC_SFC_IO_1_SDIO_D3,
    IOT_IO_FUNC_SFC_IO_1_GPIO10,
    IOT_IO_FUNC_SFC_IO_1_SPI0_CK = 4,
} IotIoFuncSfcIo1;

/*
 * @ingroup iot_io
 *
 * SFC_WPN 管脚功能
 */
typedef enum {
    IOT_IO_FUNC_SFC_IO_2_SFC_WPN,
    IOT_IO_FUNC_SFC_IO_2_SDIO_CMD,
    IOT_IO_FUNC_SFC_IO_2_GPIO11,
    IOT_IO_FUNC_SFC_IO_2_RF_TX_EN_EXT,
    IOT_IO_FUNC_SFC_IO_2_SPI0_RXD,
} IotIoFuncSfcIo2;

/*
 * @ingroup iot_io
 *
 * SFC_DI 管脚功能
 */
typedef enum {
    IOT_IO_FUNC_SFC_IO_0_SFC_DI,
    IOT_IO_FUNC_SFC_IO_0_SDIO_CLK,
    IOT_IO_FUNC_SFC_IO_0_GPIO12,
    IOT_IO_FUNC_SFC_IO_0_RF_RX_EN_EXT,
    IOT_IO_FUNC_SFC_IO_0_SPI0_CSN,
} IotIoFuncSfcIo0;

/*
 * @ingroup iot_io
 *
 * SFC_CLK 管脚功能
 */
```

```c
typedef enum {
    IOT_IO_FUNC_SFC_CLK_SFC_CLK,
    IOT_IO_FUNC_SFC_CLK_SDIO_D0,
    IOT_IO_FUNC_SFC_CLK_GPIO13,
    IOT_IO_FUNC_SFC_CLK_SSI_DATA = 4,
} IotIoFuncSfcClk;

/*
 * @ingroup iot_io
 *
 * SFC_HOLDN 管脚功能
 */
typedef enum {
    IOT_IO_FUNC_SFC_IO_3_SFC_HOLDN,
    IOT_IO_FUNC_SFC_IO_3_SDIO_D1,
    IOT_IO_FUNC_SFC_IO_3_GPIO14,
    IOT_IO_FUNC_SFC_IO_3_SSI_CLK = 4,
} IotIoFuncSfcIo3;

/*
 * @ingroup iot_io
 *
 * IO 驱动能力
 * 注意: HI_IO_NAME_GPIO_0~HI_IO_NAME_GPIO_11、HI_IO_NAME_GPIO_13~HI_IO_NAME_GPIO_14 驱
 * 动能力可选范围是 HI_IO_DRIVER_STRENGTH_0~HI_IO_DRIVER_STRENGTH_3,其余 IO 范围是 HI_IO_
 * DRIVER_STRENGTH_0~HI_IO_DRIVER_STRENGTH_7
 */
typedef enum {
    IOT_IO_DRIVER_STRENGTH_0 = 0,         /* 驱动能力 0 级,驱动能力最高 */
    IOT_IO_DRIVER_STRENGTH_1,             /* 驱动能力 1 级 */
    IOT_IO_DRIVER_STRENGTH_2,             /* 驱动能力 2 级 */
    IOT_IO_DRIVER_STRENGTH_3,             /* 驱动能力 3 级 */
    IOT_IO_DRIVER_STRENGTH_4,             /* 驱动能力 4 级 */
    IOT_IO_DRIVER_STRENGTH_5,             /* 驱动能力 5 级 */
    IOT_IO_DRIVER_STRENGTH_6,             /* 驱动能力 6 级 */
    IOT_IO_DRIVER_STRENGTH_7,             /* 驱动能力 7 级,驱动能力最低 */
    IOT_IO_DRIVER_STRENGTH_MAX,
} IotIoDriverStrength;


unsigned int IoSetPull(unsigned int id, IotIoPull val);


unsigned int IoSetFunc(unsigned int id, unsigned char val);

unsigned int task_msleep(unsigned int ms);

#endif
```

B.2 wifiiot_gpio_ex.c

```c
#include "iot_errno.h"
#include "iot_gpio_ex.h"
#include "hi_gpio.h"
#include "hi_io.h"
#include "hi_task.h"
#include "hi_types_base.h"
#include "hi_types_base.h"


unsigned int IoSetPull(unsigned int id, IotIoPull val)
{
    if (id >= HI_GPIO_IDX_MAX) {
        return IOT_FAILURE;
    }
    return hi_io_set_pull((hi_io_name)id, (hi_io_pull)val);
}

unsigned int IoSetFunc(unsigned int id, unsigned char val)
{
    if (id >= HI_GPIO_IDX_MAX) {
        return IOT_FAILURE;
    }
    return hi_io_set_func((hi_io_name)id, val);
}

unsigned int task_msleep(unsigned int ms)
{
    if (ms <= 0) {
        return IOT_FAILURE;
    }
    return hi_sleep((hi_u32)ms);
}
```

B.3 BUILD.gn

```
static_library("hal_iothardware") {
    if (board_name == "hispark_pegasus") {
        sources = [
            "hal_iot_flash.c",
            "hal_iot_gpio.c",
            "hal_iot_gpio_ex.c",
            "hal_iot_i2c.c",
            "hal_iot_pwm.c",
            "hal_iot_uart.c",
```

```
            "hal_iot_watchdog.c",
            "hal_lowpower.c",
            "hal_reset.c",
        ]
        include_dirs = [
            "//utils/native/lite/include",
            "//base/iot_hardware/peripheral/interfaces/kits",
            "//device/hisilicon/hispark_pegasus/sdk_liteos/include",
        ]
    }
}
```

附录C 系统编译与构建

C.1 Ninja 系统

Ninja 是一个小型构建系统,类似 make,其由更高级别的构建系统生成其输入文件,并且它被设计为尽可能快地运行构建。Ninja 构建文件是可读的,但手工编写并不是特别方便。

Ninja 文件没有分支、循环的流程控制,比 Makefile 简单。

Ninja 文件中通过 rule 定义规则,描述构建中需要执行的动作,由一个命令(command)和描述(description)参数组成。规则为命令行,声明一个简短的名称。它们由关键字 rule 和一个规则名称开头的行开始,然后紧跟着一组带缩进格式的 variable = value 行组成。下面示例中声明了一个名为 cc 的 rule。

```
cflags = -Wall

rule cc
 command = gcc $cflags -c $in -o $out

build foo.o: cc foo.c
```

Ninja 支持为字符串声明更短的可重用名称,如下声明了 cflags。

```
cflags = -g
```

可以用在等号的右侧,用美元符号取消引用,如下。

```
rule cc
 command = gcc $cflags -c $in -o $out
```

也可以使用花括号来引用变量,例如 ${in}。

Ninja 文件中通过 build 语句描述构建中输入和输出文件的关系。构建语句由关键字 build 开头,格式为

```
build outputs: rulename inputs
```

这样的一个声明,所有的输出文件来源于输入文件。当缺少输出文件或变更输入文件时,Ninja 将会运行此规则来重新生成输出,如 build foo.o cc foo.c 指明 foo.o 由 foo.c 通过规则 cc 而构建。在 build 块的范围内(包括对其关联的评估 rule),变量 $in 表示输入列表,变量 $out 表示输出列表。

一个 build 语句后面可以跟一组缩进的 key = value 对。在执行命令中的变量时,这些变量将隐藏任何变量。例如:

```
cflags = -Wall -Werror
rule cc
 command = gcc $cflags -c $in -o $out

#如果未指定,build 获取外部 $cflags
build foo.o: cc foo.c

#但是可以为特定 build 隐藏变量 $cflags
build special.o: cc special.c
 cflags = -Wall

#该变量仅在 special.o 的范围内被隐藏
#后续构建行获取外部(原始)cflags
build bar.o: cc bar.c
```

安装 Ninja 的过程参考 1.4.1 节,安装 Ninja 之后,通过 ninja -help 命令查看帮助,如下。

```
-version          #打印版本信息
-v, --verbose     #显示构建中的所有命令行
-C DIR            #在执行操作之前,切换到 dir 目录
-f FILE           #指定构建输入文件。默认为当前目录下的 build.ninja
-j N              #并行执行 N 个作业。默认 N = 6(和系统有关)
-k N              #持续构建直到 N 个作业失败为止。默认 N = 1
-l N              #如果平均负载大于 N,不启动新的作业
-n                #排练(dry run)(不执行命令,视其成功执行)
-d MODE           #开启调试(用 -d list 罗列所有的模式)
-t TOOL           #执行一个子工具(用 -t list 列出所有子命令工具)
-w FLAG           #调整告警级别
```

通过下面的示例熟悉 Ninja 的使用。

创建一个头文件 test.h 和源文件 test.c,其代码如下。

```
//test.h
#ifndef TEST_H
#define TEST_H
#define CONST 2022
#endif

//test.c
#include <stdio.h>
#include "test.h"
int main()
```

```
        {
                printf("Hello World!\n");
                printf("CONST =  % d\n",CONST);
                return 0;
        }
```

在同一目录下创建 build.ninja 文件,文件内容如下。

```
rule dp
 command = touch $ out
rule CC
 command = gcc - C $ in - o $ out
rule lk
 command = gcc $ in - o $ out

build test.c : dp test.h
build test.o : CC test.c
build test.out : lk test.o

default test.out
```

打开命令行定位到源码目录,执行 ninja→log.txt。log 中包含的项有开始时间、结束时间、mtime、output 文件路径名和命令行 hash。

C.2　gn 系统

gn 是一个生成 Ninja 构建文件的元构建系统,以便可以用 Ninja 构建项目。gn 把.gn 文件转换成.ninja 文件,然后 Ninja 根据.ninja 文件将源码生成目标程序。安装 gn 的过程参考 1.4.1 节,运行 gn 时,从命令行运行 gn 即可。

在命令行使用 gn help 命令能展示 gn 的各个命令,如图 C.1 所示。

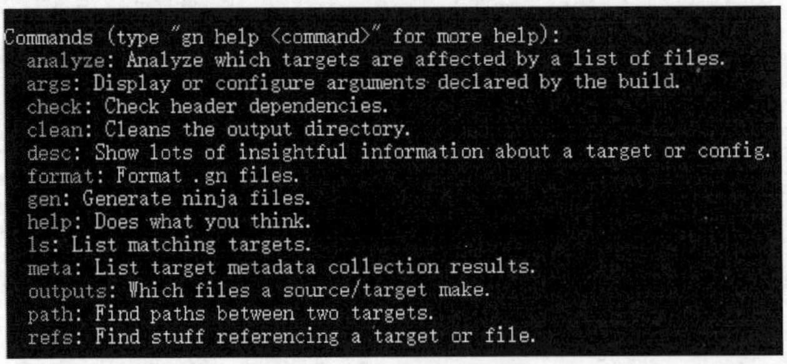

```
Commands (type "gn help <command>" for more help):
  analyze: Analyze which targets are affected by a list of files.
  args: Display or configure arguments declared by the build.
  check: Check header dependencies.
  clean: Cleans the output directory.
  desc: Show lots of insightful information about a target or config.
  format: Format .gn files.
  gen: Generate ninja files.
  help: Does what you think.
  ls: List matching targets.
  meta: List target metadata collection results.
  outputs: Which files a source/target make.
  path: Find paths between two targets.
  refs: Find stuff referencing a target or file.
```

图 C.1　gn 的命令

以下示例了一些命令的使用方法。

```
gn gen out/dir [ -- args = "..."]
        创建新的编译目录,会自动创建 args.gn 文件作为编译参数
        gn args -- list out/dir
        列出可选的编译参数
        gn ls out/dir
```

列出所有的 target
gn ls out/dir "//: hello_world*"
列出匹配的 target
gn desc out/dir "//: hello_world"
查看指定 target 的描述信息，包括 src 源码文件、依赖的 lib、编译选项等
gn refs out/dir 文件
查看依赖该文件的 target
gn refs out/dir //: hello_world
查看依赖该 target 的 target

在项目中使用 gn，必须遵循以下要求。

(1) 在根目录创建.gn 文件，该文件用于指定 BUILDCONFIG.gn 文件的位置。
(2) 在 BUILDCONFIG.gn 中指定编译时使用的编译工具链。
(3) 在独立的 gn 文件中定义编译使用的工具链。
(4) 在项目根目录下创建 BUILD.gn 文件，指定编译的目标。

根文件.gn 是一个入口设置的文件，是 gn 中的固定规则文件，会自动被 gn 读取。该文件没有文件名，只有扩展名。输入以下命令查看示例的.gn 文件模板。

```
gn help dotfile
```

默认.gn 文件模板内容如下。

```
buildconfig = "//build/config/BUILDCONFIG.gn"

        check_targets = [
        "//doom_melon/*",                    #检查该子树
        "//tools: mind_controlling_ant",     #检查指定目标
        ]

        root = "//: root"

        secondary_source = "//build/config/temporary_buildfiles/"

        default_args = {
                #该项目默认释放构建
                is_debug = false
                is_component_build = false
        }
```

在项目的根目录下创建 build/config/目录和目录中的 BUILDCONFIG.gn 文件。一个 BUILDCONFIG.gn 例子文件内容如下。

```
#Windows 工具链
        set_default_toolchain("//gn/toolchain: msvc")
        default_toolchain_name = "msvc"
        host_toolchain = "msvc"
```

在 gn 目录下创建 toolchain 目录，里面创建 BUILD.gn 文件。工具链是一组构建命令来运行不同类型的输入文件和链接的任务，具体内容参考 Google 官网。

src 目录里，创建一个编译文件 BUILD.gn，内容如下。

```
executable("hello_world") {
            # source
            sources = [
             "test.c",
            ]
        }
```

同时,该目录下创建一个源文件 test.c。文件内容如下。

```c
#include <stdio.h>

  int main()
  {
      printf("Hello World!\n");
      printf("CONST = %d\n",3);
      return 0;
  }
```

根目录下创建 BUILD.gn 文件,新创建的编译文件添加到其中一个根 group 的依赖项中(这里的每个 group 是其他目标的一个集合)。内容如下。

```
group("root") {
            deps = [
            "//src: hello_world",
            ]
        }
```

目标文件标签的前面是"//"符号(表示源码的根目录),后面紧跟具体路径,再接一个冒号,最后是项目的目标名称。

最后可以直接用以下命令执行编译。

```
gn gen out/Default
  ninja -C out/Default
```

1. 语法

gn 使用了简单的动态类型语言,支持的数据类型有布尔型、64 位有符号整数、字符串、列表和作用域。

字符串用双引号括起来,并使用反斜杠作为转义字符。唯一支持的转义序列如下。

\"(用于直接应用)
\$(字面上的美元符号)
\\(用于文字反斜杠)

任何其他反斜杠的使用都被视为文字反斜杠。所以,例如,\b 在模式中使用不需要转义,大多数 Windows 系统路径也不需要。

使用 $ 支持简单的变量替换,其中美元符号后的单词被替换为变量的值。如果没有非变量名字符来终止变量名,可以选择{}包围变量名。更复杂的表达式不被支持,仅支持变量名的替换。

```
a = "mypath"
```

```
b = "$a/foo.cc"           #b -> "mypath/foo.cc"
c = "foo${a}bar.cc"       #c -> "foomypathbar.cc"
```

列表类型支持将一个列表追加到另一个列表。追加的是项目,而不是将列表追加为嵌套成员。

```
a = [ "first" ]
    a += [ "second" ]            #[ "first", "second" ]
    a += [ "third", "fourth" ]   #[ "first", "second", "third", "fourth" ]
    b = a + [ "fifth" ]          #[ "first", "second", "third", "fourth", "fifth" ]
```

列表支持用下标获取值,下标从 0 开始。

```
a = [ "first", "second", "third" ]
    b = a[1]    # -> "second"
```

获取下标的运算符是只读的,不能用来改变列表。将非空列表分配给包含现有非空列表的变量是错误的,如下分别示意了错误和正确的书写。

```
a = ["one"]
    a = ["two"]                              #用非空列表覆盖非空列表,错误
    a = []
    a = ["two"]                              #正确
```

列表中使用"-"运算符删除项目。示例代码如下。

```
a = [ "first", "second", "third", "first" ]
    b = a - [ "first" ]                      #[ "second", "third" ]
    a -= [ "second" ]                        #[ "first", "third", "fourth" ]
```

gn 文件中可以使用条件判断语句,示例如下。

```
if(is_linux ||(is_win && target_cpu == "x86")){
        sources -= [ "something.cc" ]
    }else if(...){
        ...
    }else {
        ...
    }
```

函数可以接受用{}包围的代码块,示例代码如下。

```
static_library("mylibrary"){
        sources = ["a.cc"]
    }
```

文件和函数调用后面跟着{ }块引入新的作用域。作用域是嵌套的。当读取一个变量时,将会以相反的顺序搜索包含的作用域,直到找到匹配的名称。变量写入总是进入最内层的作用域。

除了最内层的作用域以外,没有办法修改任何封闭作用域。这意味着当定义一个目标时,例如,在块内部做的任何事情都不会泄露到文件的其余部分。

2. 静态库

如果有静态库,gn 中需要声明依赖。

构建一个有一个函数的静态库,功能是对任何人说 hello。该函数包含在一个 hello.c 的文件中。打开 src/BUILD.gn,将下面的配置静态库的文本添加到该文件的最下面。

```
static_library("hello") {
        sources = [
            "hello.cc",
        ]
}
```

添加一个可执行文件的工程,依赖上面的静态库"hello"。

```
executable("say_hello") {
        sources = [
            "say_hello.c",
        ]
        deps = [
            ":hello"
        ]
}
```

这个 exe 的工程包含一个源文件并且依赖上面的 lib 工程。lib 是通过 exe 中的 deps 依赖项引入的。可以使用全路径文本//src:hello,但是如果引入的工程是在同一个 BUILD.gn 文件中,可以使用缩写:hello。

在源码根目录中输入以下命令测试静态库。

```
ninja -C out/Default say_hello
```

不需要再次运行 gn 命令。当 BUILD.gn 文件改变时 gn 会自动重新生成 Ninja 文件。这通常发生在 Ninja 刚运行时,它会打印出"[1/1] Regenerating ninja files"。

3. 预编译设置

hello 库工程有一个新功能,是可以同时对两个人说 hello。这个功能是通过宏定义 TWO_PEOPLE 来控制的。在工程中添加以下宏定义。

```
static_library("hello") {
        sources = [
            "hello.c",
        ]
        defines = [
            "TWO_PEOPLE",
        ]
}
```

然而,使用该工程也需要知道这个预定义,将定义放到库工程中,仅对这个工程中的文件生效。如果别的工程包含 hello.h,这些工程是看不到这个定义的。如果要看到这个定义,每个引入这个库的工程都必须定义 TWO_PEOPLE。

4. 新增配置

gn 有一个"config"的概念,可以将一些设置放在里面。如下所示创建一个"config"并将需要的预定义宏放到里面。

```
config("hello_config") {
    defines = [
        "TWO_PEOPLE",
    ]
}
```

如何在目标工程中引入这些设置,只需要在目标的 configs 选项中将需要的配置罗列出来,如下。

```
static_library("hello") {
    sources = [
        "hello.c",
    ]
    config += [
        ":hello_config",
    ]
}
```

config 这里使用+=,而不是=,因为每个目标工程构建编译的都有一系列的默认配置。+=是将新的配置添加到默认的配置中,而不是全部重写它。

上面介绍的方法可以很好地封装配置,但是它仍然需要每个使用库的工程在它们自己的配置里去设置。通过以下写法,每个使用 lib 的工程可以自动获取这些配置。现在来改变以下库的配置。

```
static_library("hello") {
    sources = [
        "hello.c",
    ]
    all_dependent_configs = [
        ":hello_config",
    ]
}
```

这会将 hello_config 的配置应用到 hello 工程本身,以及所有依赖(所有继承,包含间接依赖)hello 的工程。现在每个依赖库的工程都会获取这些配置。也可以设置 public_configs 来设置只应用到直接依赖的目标项目中(不传递)。

config 可以提供一个公共的配置对象,除了包括编译 defines,也可以是 flag 和 include 等,可被其他 target 包含,如下。

```
config("myconfig") {
    include_dirs = [ "include/common" ]
    defines = [ "ENABLE_DOOM_MELON" ]
}

executable("mything") {
    configs = [ ":myconfig" ]
}
```

构建配置文件通常指定设置默认配置列表的目标默认值。目标可以根据需要添加或删除。所以通常会使用 configs ＋＝ ":myconfig"追加到默认列表。

5. 新增编译参数

可以通过 declare_args 直接声明所接受的参数并指定它们的默认值，如下。

```
declare_args() {
    enable_test = true
    enable_doom_melon = false
}
```

新增的 enable_test 的 gn 编译参数，默认值是 true。在 BUILD.gn 文件中，根据这个编译参数的值进行一些特殊化配置，如下。

```
if(enable_test)
    {
        …
    }
```

如果要了解编译参数详情，使用 gn help buildargs 命令。使用 gn help declare_args 获取声明参数的细节。

6. 新增编译单元

target 是一个最小的编译单元，可以将它单独传递给 Ninja 进行编译。Google 文档有以下几种 target。

（1）action：Declare a target that runs a script a single time. 声明运行一次脚本的 target。

（2）action_foreach：Declare a target that runs a script over a set of files. 声明运行一组文件的脚本的 target。

（3）bundle_data：[iOS/macOS]Declare a target without output. 声明一个无输出文件的 target。

（4）copy：Declare a target that copies files. 声明复制文件的 target。

（5）create_bundle：[iOS/macOS]Build an iOS or macOS bundle. 编译 MACOS/IOS 包。

（6）executable：Declare an executable target. 生成可执行 target。

（7）generated_file：Declare a generated_file target. 声明 generated_file target。

（8）group：Declare a named group of targets. 声明有组名的 target 组。

（9）loadable_module：Declare a loadable module target. 声明可加载的模块目标。

（10）rust_library：Declare a Rust library target. 声明 Rust 库目标。

（11）shared_library：Declare a shared library target. 声明动态链接库，.dll 或 .so。

（12）source_set：Declare a source set target. 声明源集 target，比 static_library 要快。

（13）static_library：Declare a static library target. 声明静态库目标，.lib 或 .a。

（14）target：Declare an target with the given programmatic type. 用给定的编程类型声明 target。

7. 新增模板

模板是可以用来定义可重用的代码，比如添加新的 target 类型等。通常可以将模板单独定义成一个 .gni 文件，然后其他文件就可以通过 import 来引入实现共享。

8. 新增依赖关系

在编译的时候需要小心地处理各种动态库和静态库的链接引入，在 gn 中，使用 deps 来实现库的依赖关系，如下。

```
if (enable_nacl) {
        deps += [ "//components/nacl/loader: nacl_loader_unittests" ]

        if (is_linux) {
                deps += [ "//components/nacl/loader: nacl_helper" ]

                if (enable_nacl_nonsfi) {
                        deps += [
                        "//components/nacl/loader: helper_nonsfi",
                        "//components/nacl/loader: nacl_helper_nonsfi_unittests",
                        ]
                }
        }
}
```

C.3 轻量级系统编译构建

OpenHarmony 构建系统应用了 gn 和 Ninja，支持 OpenHarmony 的组件化开发。OpenHarmony 的组件是系统最小的可复用、可配置、可裁剪的功能单元。组件具备目录独立可并行开发、可独立编译、可独立测试的特征。一个或多个具体的组件组成子系统。OpenHarmony 整体遵从分层设计，从下向上依次为内核层、系统服务层、框架层和应用层。系统功能按照"系统→子系统→组件"逐级展开，在多设备部署场景下，支持根据实际需求裁剪某些非必要的子系统或组件。

OpenHarmony 编译构建过程如图 C.2 所示。

其中，hb 是 OpenHarmony 的命令行工具，用来执行编译命令。OpenHarmony 编译构建主要分为设置和编译两步。

（1）应用 hb set 设置 OpenHarmony 源码目录和要编译的产品。

（2）应用 hb build 编译产品、开发板或者组件。编译过程如下。

① 读取编译配置。根据产品选择的开发板，读取开发板 config.gni 文件内容，主要包括编译工具链、编译连接命令和选项等。

② 调用 gn。调用 gn gen 命令，读取产品配置生成产品解决方案 out 目录和 Ninja 文件。

③ 调用 Ninja。调用 ninja -C out/board/product 启动编译。

④ 系统镜像打包。将组件编译产物打包，设置文件属性和权限，制作文件系统镜像。

为了实现芯片解决方案、产品解决方案与 OpenHarmony 是解耦的、可插拔的，

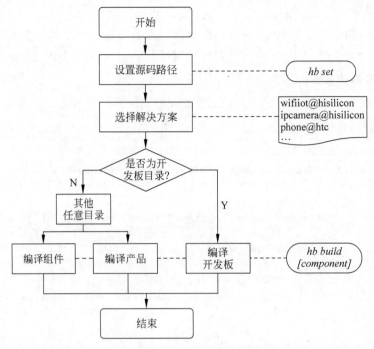

图 C.2　OpenHarmony 编译构建过程

OpenHarmony 的组件、芯片解决方案和产品解决方案的路径、目录树和配置需遵循一定的规则，这里仅对组件的配置做一简介。

组件源码路径命名规则为领域/子系统/组件。组件目录树规则如图 C.3 所示。

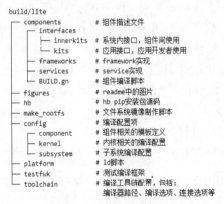

图 C.3　组件目录树

组件的名称、源码路径、功能简介、是否必选、编译目标、RAM、ROM、编译输出、已适配的内核、可配置的特性和依赖等属性定义在 build/lite/components 目录下对应子系统的 JSON 文件中，新增组件时需要在对应子系统 JSON 文件中添加相应的组件定义。产品所配置的组件必须在某个子系统中被定义过，否则会校验失败。一个 sensor 服务组件的属性定义描述文件字段如下。

{

```
"components": [
{
        "component": "sensor_lite",            #组件名称
        "description": "Sensor services",      #组件一句话功能描述
        "optional": "true",                    #组件是否为最小系统必选
        "dirs": [                              #组件源码路径
        "base/sensors/sensor_lite"
        ],
        "targets": [                           #组件编译入口
        "//base/sensors/sensor_lite/services: sensor_service"
        ],
        "rom": "92KB",                         #组件 ROM 值
        "ram": "~200KB",                       #组件 RAM 估值
        "output": [ "libsensor_frameworks.so" ], #组件编译输出
        "adapted_kernel": [ "liteos_a" ],      #组件已适配的内核
        "features": [],                        #组件可配置的特性
        "deps": {
            "components": [                    #组件依赖的其他组件
            "samgr_lite",
            "ipc_lite"
            ],
            "third_party": [                   #组件依赖的三方开源软件
            "bounds_checking_function"
            ]
        }
}
]
}
```

在编写组件构建文件 BUILD.gn 时，OpenHarmony 建议编译目标名称与组件名称一致。组件对外可配置的特性变量需声明在该组件 BUILD.gn 中，特性变量命名规则为 ohos_subsystemcomponentfeature。特性在组件描述中也需要同步定义，在产品配置文件 config.json 中按需配置。宏定义规则为 OHOS_SUBSYSTEMCOMPONENTFEATURE。

以图形的 UI 组件为例，foundation/graphic/ui/BUILD.gn 文件如下。

```
#声明组件可配置的特性
    declare_args() {
        enable_ohos_graphic_ui_animator = false   #动效特性开关
        ohos_ohos_graphic_ui_font = "vector"      #可配置的字体类型,vector 或者 bitmap
    }

    #组件基础功能
    shared_library("base") {
        sources = [
        ...
        ]
        include_dirs = [
        ...
        ]
```

```
    }

    # 仅在 animator 开启时编译
    if(enable_ohos_graphic_ui_animator ) {
        shared_library("animator") {
            sources = [
            ...
            ]
            include_dirs = [
            ...
            ]
            deps = [ : base ]
        }
    }
    ...
    # target 名称建议与组件名称一致,组件 target 类型可以是 executable(bin 文件)、shared_
library(动态库.so)、static_library(静态库.a)、group 等
    executable("ui") {
        deps = [
        ": base"
        ]

        # animator 特性由产品配置
        if(enable_ohos_graphic_ui_animator ) {
            deps += [
            "animator"
            ]
        }
    }
```

后　　记

在写下这本书的第一个字之前，有一句广告语，经常浮现在我的脑海中，这句广告语就是，"我们不生产某某，我们是大自然的搬运工。"相对于其他同行企业，这个搬运工企业的毛利率远远大于同行。这颠覆了通常对一般搬运工低收入的认识。在全国闻名的搬运工系列中，无论是重庆的棒棒还是武汉的扁担，收入都不很理想，和"大自然的搬运工"相比，一个是地下，一个是天上。不过，可能存在着比大自然的搬运工利润更高的搬运工，那就是在一些文艺作品中被戏称为"他人财物的搬运工"。

我一直在思考孔老夫子的一句名言，"师者，所以传道、授业、解惑也！"孔圣人对"师"的功能进行定义时，把"传道"排在首位，而"传道"，我想，应该具有类似搬运的含义。作为老师，某种意义上是不是也可以说，"老师不生产知识，只是他人知识的搬运工。"这个说法似乎有点儿道理，也有点儿合适。本书可以看做是一个搬运工，搬运知识、搬运技术、搬运道理、搬运文字。

不过，摆在面前的一个问题是本书要搬运哪些知识？自2018年以来，针对中国企业实体清单越来越多。2018年4月16日，美国商务部工业和安全局禁止美国企业向中兴公司出售零部件产品，期限为7年。2019年5月15日，美国商务部工业和安全局将华为公司列入了实体清单进行制裁，实行对华为的技术封锁，禁止了华为5G技术在美国市场上的准入。为了阻止华为的发展，美国一再修改其对华为的禁令进行技术封锁。从2020年5月15日禁止华为使用美国芯片设计软件，到2020年8月17日禁止含有美国技术的代工企业生产芯片给华为，再到2020年9月15日禁止拥有美国技术成分的芯片出口给华为。自此美国对华为的芯片管制令正式生效，台积电、高通、三星、中芯国际等多家公司将不能再供应芯片给华为。随之而来的是华为产品的国际市场份额急剧下降。但是华为没有被困难吓倒，没有在险阻面前后退，更没有屈服，体现了中华民族坚韧不拔的优良品质。

处于这样的国际环境之中，选择搬运华为的技术和知识是必需的。不过，在当前国际化大趋势之下，很难说一项技术没有前人或者他人的影子，或者完全属于某人或者某个企业或者某个国家。在华为的各项技术中，其力推的鸿蒙操作系统一直受到较高的关注度，因此本书选择了搬运鸿蒙操作系统的相关技术。鸿蒙操作系统从操作系统层面上打通了多态、多设备的连接，功能和内容巨多。通过一本书涵盖整个系统的技术是不可能的，本书仅有选择性地搬运了物联网应用方面的一些基础知识。对于鸿蒙操作系统中更有吸引力的、更深入的技术，本书由于篇幅限制没有过多涉及，不得不说是个遗憾了。

希望本书所搬运之物能够为需要者提供有益的帮助。在本书成稿期间，部分内容已经帮助学生团队在华为ICT学院创新赛中获得了好成绩。

搬运工并不负责物品的真假、优劣，搬运过程中可能出现一些遗漏，期望读者不吝指教，在此先表示诚挚的谢意。

葛　非

2022年11月